SpringerBriefs in Education

For further volumes:
http://www.springer.com/series/8914

Mansoor Niaz · Arelys Maza

Nature of Science in General Chemistry Textbooks

 Springer

Prof. Mansoor Niaz
Epistemology of Science Group
Department of Chemistry
Universidad de Oriente
Apartado Postal 90
6101 Cumaná, Estado Sucre
Venezuela
e-mail: niazma@gmail.com

Prof. Arelys Maza
Epistemology of Science Group
Department of Chemistry
Universidad de Oriente
Apartado Postal 90
6101 Cumaná, Estado Sucre
Venezuela
e-mail: arelmazw@hotmail.com

ISSN 2211-1921
ISBN 978-94-007-1919-4
DOI 10.1007/978-94-007-1920-0
Springer Dordrecht Heidelberg London New York

e-ISSN 2211-193X
e-ISBN 978-94-007-1920-0

Cover design: eStudio Calamar, Berlin/Figueres

Printed on acid-free paper

Springer is part of Springer Science+Business Media (www.springer.com)

Acknowledgments

Research reported here was supported in part by various grants from Consejo de Investigación, Universidad de Oriente (Venezuela).

We would like to express our sincere thanks to the following members of our research group who participated in various stages of research by suggesting improvements that helped to clarify underlying issues related to nature of science: Ysmandi Páez, and Luis A. Montes. We have also benefitted from discussions and criticisms at different stages from: Fouad Abd-El-Khalick (University of Illinois at Urbana-Champaign), Stephen Klassen (University of Winnipeg) and Liberato Cardellini (Università Politecnica delle Marche, Italy). The two anonymous reviewers provided constructive criticisms, which helped to improve the monograph.

A special word of thanks is due to Bernadette Ohmer, Publishing Editor at Springer (Dordrecht) for her support and encouragement throughout the different stages preceding publication.

Contents

Chapter 1
Nature of Science in General Chemistry Textbooks

Abstract Research in science education has recognized the importance of nature of science (NOS) within history and philosophy of science (HPS) perspective. The objective of this study is to evaluate representation of NOS in 75 general chemistry textbooks (published in USA) based on nine criteria. Depending on the treatment of the criteria, textbooks were classified as: No mention (N), Mention (M), and Satisfactory (S). Most textbooks in this study provided little insight with respect to the nine criteria used for evaluating NOS. Percentage of textbooks that were classified as No mention (N) ranged from 44 (Criterion 1) to 94.7% (Criterion 8). Despite this, some textbooks provided good examples based on HPS, and the percentage of textbooks that were classified as Satisfactory (S) ranged from 1.3 (Criterion 2) to 17.3% (Criterion 1). These examples show that although presentation of NOS is not the major objective of general chemistry textbooks, some of them inevitably refer to the historical record and thus provide guidelines for future textbooks that align with the reform documents. Some textbooks go into considerable detail to present the atomic models of Dalton, Thomson, Rutherford, Bohr and wave-mechanical. However, the most important aspect of these presentations is that they explicitly do so in the context of the tentative nature of scientific theories (Criterion 1). This is a clear illustration of how the history of chemistry can facilitate the understanding of NOS. Similar evidence based on various historical episodes on other criteria is reported. It is concluded that in most cases the history of chemistry is 'inside' chemistry and in order to facilitate understanding, textbooks need to interpret within a NOS perspective.

Keywords Science education · Nature of science · General chemistry textbooks · History and philosophy of science · Science curriculum · Dynamics of scientific progress · Tentative nature of scientific theories · Scientific method · Observations are theory-laden · Rational arguments, creativity and skepticism · Competition between rival theories · Inconsistent foundations · Atomic models · Hierarchical relation between laws and theories · Objective nature of science · Social and

M. Niaz and A. Maza, *Nature of Science in General Chemistry Textbooks*,
SpringerBriefs in Education, DOI: 10.1007/978-94-007-1920-0_1,
© Mansoor Niaz 2011

historical milieu · Alternative interpretations · Suspension of disbelief · History of chemistry is 'inside' chemistry · Scientific laws as idealizations · Quantitative imperative · Imperative of presuppositions · Teaching science as practiced by scientists · Author or publisher effect

1.1 Introduction

Research in science education has recognized the importance of history and philosophy of science (HPS). Similarly, reform documents in various parts of the world have also espoused the inclusion of HPS in both the science curriculum and in textbooks, in order to facilitate a vision of science more in consonance with progress in science (American Association for the Advancement of Science, AAAS 1993; Project Beyond 2000 in the UK, Millar and Osborne 1998). More recently, Niaz (2010) has argued that in order to familiarize students and teachers with the dynamics of scientific progress, we need to teach science as practiced by scientists. Among other aspects, most researchers would agree that nature of science (NOS) is an essential part of HPS with important implications for teaching science (Hodson 2009). Furthermore, research also shows that most high school and freshman students (and even teachers) in many parts of the world have NOS views that are quite close to an empiricist epistemology and do not align with the reform documents and recent HPS-based educational research (Abd-El-Khalick 2005; Akerson et al. 2010; Blanco and Niaz 1997; Cobern et al. 1999; Dagher and BouJaoude 1997; Dogan and Abd-El-Khalick 2008; Irez 2006; Kang et al. 2005; Khishfe 2008; Lederman 1992; Niaz 2008a; Sadler et al. 2004; Tsai 2007).

The role played by textbooks in developing students' informed NOS conceptions has been a source of particular concern for the reform documents (Stern and Roseman 2004). In many parts of the world, the textbook is the only resource for the teacher and to make matters worse, textbooks become *the* curriculum and determine to a great extent what is taught and learned in the classroom (Abd-El-Khalick et al. 2008). Recent HPS-based research has shown increasing interest in analyzing textbooks and thus providing guidelines for future textbooks. Some of the topics that have been researched are the following:

(a) Scientific literacy themes (Chiappetta et al. 1991; Leite 2002).
(b) Normal science education (Van Berkel et al. 2000).
(c) Atomic structure (Justi and Gilbert 2000; Niaz 1998; Rodríguez and Niaz 2002, 2004a; Niaz and Coştu2009).
(d) Determination of the elementary electrical charge (Niaz 2000a; Rodríguez and Niaz 2004b; Niaz and Rodríguez 2005).
(e) Amount of substance and its unit the 'mole' (Padilla and Furio-Mas 2008).
(f) Laws of definite and multiple proportions in chemistry (Niaz 2001a).

(g)   Periodic table of chemical elements (Brito et al. 2005).
(h)   Osmotic pressure (De Berg 2006).
(i)   Heat and temperature (De Berg 2008).
(j)   Quantum mechanics (Niaz and Fernández 2008; Shiland 1997; Tampakis and Skordoulis 2007).
(k)   Relativity theory (Arriassecq and Greca 2007; Velentzas et al. 2007).
(l)   The pendulum (Koliopoulos and Constantinou 2005).
(m)   Photoelectric effect (Niaz et al. 2010a).
(n)   Evolution (Skoog 2005).

Besides these particular topics of the science curriculum, science education research has also expressed interest in the underlying common denominator of these topics, namely NOS (Abd-El-Khalick et al. 2008; Chiappetta and Fillman 2007; Guisasola et al. 2005; Irez 2009; Koul and Dana 1997; Páez and Niaz 2008). Abd-El-Khalick et al. (2008) have drawn attention to the importance of including NOS in high school chemistry textbooks. These authors analyzed 14 textbooks (published in USA) including five 'series' spanning one to four decades, with respect to the following NOS aspects: empirical, tentative, inferential, creative, theory-driven, myth of the scientific method, nature of scientific theories and laws, and the social and cultural embeddedness of science. Results from this study revealed that high school chemistry textbooks fared poorly in their representation of NOS, which led the authors to conclude, "These trends are incommensurate with the discourse in national and international science education reform documents ..." (p. 835).

At this stage it is important to note an important difference between this study and that of Abd-El-Khalick et al. (2008): these authors analyzed the whole textbooks as the different chapters had not been previously analyzed. Our criteria are quite similar to those of Abd-El-Khalick et al., which are derived from a fair degree of consensus in the science education research community. However, our problem situation was different as over 80% of the general chemistry (introductory university freshman level) textbooks had already been analyzed with respect to the following topics:

(a)   Atomic structure (Niaz 1998; Rodríguez and Niaz 2002).
(b)   Oil drop experiment (Niaz 2000a).
(c)   Kinetic molecular theory of gases (Niaz 2000b).
(d)   Laws of definite and multiple proportions (Niaz 2001a).
(e)   Covalent bond (Niaz 2001b).
(f)   Periodic table (Brito et al. 2005).
(g)   Quantum Numbers (Niaz and Fernández 2008).

These studies are based on a particular context of the curriculum and in a sense do not provide an overview of textbooks' philosophical orientation. Besides these studies published by our group, some other work has also been published. Thus, a fair part of the general chemistry textbooks published in USA, have already been analyzed. However, our review of the literature shows that no study has been published that analyzes the Introductory chapter (or preface) of these textbooks. Furthermore, our research experience shows that a textbook author does not necessarily present a consistent NOS perspective in all the chapters. For example, we have found that a textbook may present the chapter on atomic structure satisfactorily but does not do the same in the chapter on periodic table (see Table 1.4 for more details). On the other hand, the Introductory chapter (or preface) of a textbook does provide an overall NOS perspective of the author.

Given this background and interest, we decided to analyze and evaluate NOS in the introductory chapter of university level, general chemistry textbooks published in USA. It is plausible to suggest that the introductory chapter of the textbooks provides an overview of the authors' understanding of progress in science and hence determines the presentation of other topics of the general chemistry curriculum. Based on these considerations, this study has the following objectives:

1. Formulation of nine criteria related to NOS based on a critical evaluation of the literature.
2. Evaluation of 75 introductory, university level general chemistry textbooks (published in USA) based on NOS criteria.

1.2 Criteria for Evaluation of General Chemistry Textbooks

It is important to note that there is a fair amount of consensus in the science education literature with respect to the following NOS criteria (Abd-El-Khalick et al. 2008; Lederman et al. 2002; McComas 2008; Smith and Scharmann 2008):

1.2.1 Criterion 1: Tentative Nature of Scientific Theories

Science is not an inalterable and rigid body of 'absolute truths.' A critical appraisal of the history of science shows that scientists continually look for theories that provide greater explanatory power. At the beginning of the twentieth century, a quick succession of atomic models (Thomson, Rutherford, Bohr, Bohr–Sommerfeld, wave-mechanical) provides a good illustration of the tentative NOS. According to Lakatos (1970), theories are superseded in the degree to which a new theory provides greater heuristic/explanatory power and despite the empirical success, all scientific theories ultimately turn out to be 'false' (p. 158). Similarly, science educators have endorsed the tentative nature of scientific theories as an important characteristic of NOS (Burbules and Linn 1991; McComas et al. 1998; Niaz 2001c; Smith and Scharmann 1999).

1.2.2 Criterion 2: Laws and Theories Serve Different Roles in Science (Theories Do Not Become Laws Even with Additional Evidence)

According to the positivist/empiricist perspective of progress in science, successive verifications of a theory facilitate its conversion into a law, or vice versa, a law can be elevated to the status of a theory. Most modern philosophers of science have questioned this hierarchical/dichotomous relationship between laws and theories (Giere 1999). Based on an empiricist perspective of progress in science, most science students, teachers and even textbooks consider the relationship between theories and laws to be dichotomous/hierarchical. Mendeleev's classification of chemical elements based on atomic weights is generally considered to be an empirical law and the modern periodic table based on atomic numbers is considered to be a theoretical formulation. According to Lakatos (1970), "the clash is not 'between theories and facts' but between two high-level theories: between an *interpretative theory* to provide the facts and an *explanatory theory* to explain them; and the interpretative theory may be on quite as high a level as the explanatory theory" (p. 129, original italics). Based on this framework, Niaz et al. (2004) suggested that Mendeleev's work can be considered as an 'interpretative theory' and the periodic table based on atomic numbers as the 'explanatory theory'. In other words, scientific progress is characterized by a series of theories or models (plausible explanations), which vary in the degree to which they explain/interpret/predict the experimental findings. Researchers in science education have also questioned the dichotomy between theories and laws (McComas et al. 1998).

1.2.3 Criterion 3: There is No Universal Step-by-Step Scientific Method

The National Society for the Study of Education (1947) emphasized the scientific method for teachers in the following steps: making observations, defining the problem, constructing hypotheses, experimenting, compiling results, and drawing conclusions. This oversimplified view of what constitutes the scientific endeavor has proven to be resistant to change and is used almost in all parts of the world (Windschitl 2004). Lederman et al. (2002) have clearly traced its origin to Francis Bacon's *Novum Organum*, and its unhealthy influence on science education:

> The myth of the scientific method is regularly manifested in the belief that there is a recipe-like stepwise procedure that all scientists follow when they do science. This notion was explicitly debunked: There is no single scientific method that would guarantee the development of infallible knowledge (AAAS 1993; Bauer 1994; Feyerabend 1993; NRC 1996; Shapin 1996). It is true that scientists observe, compare, measure, test, speculate, hypothesize, create ideas and conceptual tools, and construct theories and explanations. However, there is no single sequence of activities (prescribed or otherwise)

that will unerringly lead them to functional or valid solutions or answers, let alone
certain or true knowledge (pp. 501–502).

1.2.4 Criterion 4: Observations are Theory-Laden

Scientists invariably have presuppositions and prior theoretical frameworks before
they start collecting data. At times these prior beliefs are well formulated and
resistant to change. History of science provides many examples of such frameworks and how they frequently lead to rivalries and controversies among scientists.
Determination of the elementary electrical charge provides a good example of how
two eminent scientists, Robert Millikan and Felix Ehrenhaft had two different
presuppositions to understand the same set of experimental data, namely existence
of electrons or subelectrons, respectively. Philosophers of science have emphasized the importance of such frameworks in scientific progress and refer to them in
the following terms: guiding assumptions (Laudan et al. 1988); presuppositions
(Holton 1978); and hard-core or negative heuristic of a research program (Lakatos
1970).

1.2.5 Criterion 5: Scientific Knowledge Relies Heavily,
but Not Entirely, on Observation, Experimental Evidence,
Rational Arguments, Creativity and Skepticism

History of science shows that scientists do experiments and collect data, guided by
their presuppositions. This inevitably leads them to engage in rational arguments
with their peers and colleagues. Arguments by themselves do not resolve all the
problems, as scientists resist changes in their particular way of interpreting data,
and do not abandon the hard-core of their research program as soon as anomalous
data start pouring in (Lakatos 1970). In other words scientists are skeptic of both
data and its interpretations. Understanding data is a complex and lengthy process
and requires considerable amount of ingenuity and creativity on the part of the
scientists (Holton 1978; McComas et al. 1998).

1.2.6 Criterion 6: Scientific Progress is Characterized
by Competition Between Rival Theories

Difficulties involved in scientific progress lead to rival theories and inevitably
generate controversies and conflicts. Even scientists and some philosophers
themselves have difficulties in recognizing the role played by controversies.

This dissociation between how science is actually practiced and depicted shows the need for teaching science as it is actually practiced, which facilitates a greater understanding of the dynamics of scientific progress (Niaz 2011). Machamer et al. (2000) have referred to this dilemma in succinct terms:

> Many major steps in science, probably all dramatic changes, and most of the fundamental achievements of what we now take as the advancement or progress of scientific knowledge have been controversial and have involved some dispute or another. Scientific controversies are found throughout the history of science. This is so well known that it is trivial. What is not so obvious and deserves attention is a sort of paradoxical dissociation between science as actually practiced and science as perceived or depicted by both scientists and philosophers. While nobody would deny that science in the making has been replete with controversies, the same people often depict its essence or end product as free from disputes, as the uncontroversial rational human endeavor par excellence (p. 3).

1.2.7   Criterion 7: Scientists can Interpret, the Same Experimental Data Differently

History of science shows that alternative interpretation of experimental data is one of the most interesting facets of NOS. It is generally believed that progress in science is a product of experimental data that unambiguously lead to the formulation of scientific theories. A good example is provided by alpha particle experiments conducted by the research groups of Thomson and Rutherford. Soon after Geiger and Marsden (1909) published their results (Rutherford's research group), Thomson and colleagues also started working on the scattering of alpha particles in their laboratory. Although experimental data from both laboratories were similar, interpretations of Thomson and Rutherford were entirely different. Thomson propounded the hypothesis of *compound scattering*, according to which a large angle deflection of an alpha particle resulted from successive collisions between the alpha particles and the positive charges distributed throughout the atom. Rutherford in contrast, propounded the hypothesis of *single scattering*, according to which a large angle deflection resulted from a single collision between the alpha particle and the massive positive charge in the nucleus. The rivalry led to a bitter dispute between the proponents of the two hypotheses. Rutherford even charged Crowther (1910), a colleague of Thomson, to have 'fudged' the data in order to provide support for Thomson's model of the atom (Niaz 1998; Wilson 1983). Rutherford's dilemma: On the one hand he was entirely convinced and optimistic that his model of the atom explained experimental findings better, and yet it seems that the prestige, authority, and even perhaps some reverence for his teacher made him waver. A science student may wonder as to why Thomson and Rutherford did not meet over dinner (they were well known to each other) and decide in favor of one or the other model. Progress in science is, however, much more complex. Both Thomson and Rutherford stuck to their presuppositions. Another example is provided by the oil drop experiment (Holton 1978).

1.2.8 Criterion 8: Development of Scientific Theories, at Times is Based on Inconsistent Foundations

An important aspect of Bohr's (1913) model of the atom was the presence of a deep philosophical chasm: that is, in the stationary states, the atom obeys the classical laws of Newtonian mechanics; on the other hand, when the atom emits radiation, it exhibits discontinuous (quantum) behaviour. Based on these and other arguments, Bohr's 1913 article, in general, had a fairly adverse reception in the scientific community. Lakatos (1970) has argued that Bohr employed a methodology used frequently by scientists in the past and perfectly valid for the advancement of science:

> ... some of the most important research programmes in the history of science were grafted on to older programmes with which they were blatantly inconsistent. For instance, Copernican astronomy was 'grafted' on to Aristotelian physics, Bohr's programme on to Maxwell's. Such 'grafts' are irrational for the justificationist and for the naive falsificationist, neither of whom can countenance growth on inconsistent foundations... As the young grafted programme strengthens, the peaceful co-existence comes to an end, the symbiosis becomes competitive and the champions of the new programme try to replace the old programme altogether (p. 142, original italics).

1.2.9 Criterion 9: Scientific Ideas are Affected by Their Social and Historic Milieu

Scientific knowledge is socially negotiated and this need not be confused with relativistic notions of science. According to Longino (2004):

> Establishing what the data are, what counts as acceptable reasoning, which assumptions are legitimate, and which are not become in this view a matter of social, discursive interactions as much as of interaction with the material world. Since assumptions are, by their nature, usually not explicit but taken-for-granted ways of thinking, the function of critical interaction is to make them visible, as well as to examine their metaphysical, empirical and normative implications (p. 133).

It is important to note that Abd-El-Khalick et al. (2008) consider the double-blind peer-review process used by scientific journals as one aspect of the enactment of the NOS dimensions under this aspect. Similarly, other philosophers of science have also emphasized the interaction between science, values, and objectivity (Machamer and Wolters 2004).

1.3 Evaluation of General Chemistry Textbooks: Results and Discussion

Evaluation of general chemistry textbooks (published in USA) in this study is based on an analysis of the Introductory chapter (or preface), which provides an overview of what the author(s) consider to be the essential aspects of science and how it develops. This study has evaluated NOS aspects of textbooks only in the Introductory chapter (various other chapters of these and other chemistry textbooks have already been analyzed in previous studies. Cf. Introduction section). Some of the textbooks devote considerable space (15–20pp.) to this chapter and deal with sections such as: What is chemistry;Early history of chemistry; Experimental nature of chemistry; The scientific approach to knowledge; Science and the scientific method; NOS and the scientific method; Goals, methods and characteristics of scientists. The nine criteria developed in this study are based on such issues (observations, theories, laws, scientific method, tentative nature of scientific knowledge and creativity in science) and a review of the literature in science education research. This clearly shows that the issues discussed in the Introductory chapter are quite different (nonetheless important) from the list of studies presented above (based on particular chapters of the textbooks). In a particular chapter of the textbook the criteria for evaluating textbooks are based on a detailed historical reconstruction and are more context dependent. A complete list of all the textbooks analyzed is presented in Appendix A.

1.3.1 Guidelines for Selection of Textbooks

(a) Availability of textbooks in our university and nearby libraries.
(b) Inclusion of recent textbooks.
(c) Inclusion of textbooks that have published various editions, which shows their acceptance by the science education community.
(d) Inclusion of textbooks that were published before 1990, in order to compare them with more recent textbooks. Forty-two textbooks in this study were published in the period 1965–1990.
(e) Consultations with colleagues in different parts of the world revealed that various textbooks selected for this study are used as translations.
(f) Various studies published in science education journals have used these textbooks.

1.3.2 Procedure for Applying the Criteria

The following classifications were elaborated to evaluate the textbooks:

- Satisfactory (S): Treatment of the subject in the textbook is considered to be satisfactory, if the criterion is described and examples provided to illustrate the different aspects.
- Mention (M): A simple mention of the criterion with little elaboration and no examples.
- No mention (N): No mention of the issues involved in the criterion, as conceived by this study. For example, in the case of Criterion 3, a textbook may provide a

description of the scientific method based on the traditional steps and still be classified as No mention, as it constitutes an 'uninformed' perspective.

Textbooks were awarded the following points: $S = 2$ points; $M = 1$ point; and $N = 0$ point.

1.3.2.1 Reliability of Evaluation of Textbooks Based on Inter-Rater Agreement

To begin with, two authors read and discussed various HPS-related articles and other materials in order to achieve a fair degree of consensus on critical issues. As a first step, the authors analyzed three textbooks (selected randomly) on all nine criteria. On one of the textbooks there was consensus on all nine criteria. On the second textbook there was consensus on seven criteria and on the third textbook there was consensus on five criteria. All differences were discussed and arguments presented and finally a consensus was achieved. As a next step, the authors analyzed another set of three textbooks. On one of the textbooks there was consensus on all nine criteria. On the second textbook there was consensus on eight criteria and on the third textbook there was consensus on six criteria. Once again all differences were resolved by discussion. With this experience both authors analyzed the remaining textbooks over a period of about two months. All disagreements were resolved in several meetings and in each meeting (about 2 h) the same procedure was followed. Appendix B provides complete details of the reliability of evaluation of textbooks on all nine criteria. It is important to note that the average inter-rater agreement for all nine criteria was 92.9%.

At this stage it is important to note that inter-rater estimates are essential, but nevertheless not extant from controversy. For example, in textbook analyses that depend on counting the number of figures, diagrams and tables, inter-rater agreement is high and free of disagreements. However, in the present study, evaluation of the textbooks is based on philosophical and epistemological issues that pertain to competing paradigms and hence generate considerable discussions. Taking these factors into consideration it is important that the raters must share some degree of uniformity with respect to philosophical backgrounds and fair knowledge of the relevant literature. Chiappetta et al. (2006) also raised this issue and generally agreed with the thesis of this study.

1.3.3 Criterion 1: Tentative Nature of Scientific Theories

Thirteen general chemistry textbooks (see Tables 1.1 and 1.2) were classified as Satisfactory (S), and the following are three examples:

> ... science relies upon *models* that have been constructed as the result of observations. A good model fits many separate observations and can be used to predict what will happen in a new experiment. The reputation of the model grows as predictions from it are verified.

Table 1.1 Evaluation of nature of science in general chemistry textbooks ($n = 75$)

No.	Textbook	Criteria[a]									Points[b]
		1	2	3	4	5	6	7	8	9	
1	Ander and Sonnessa (1965)	N	N	N	N	N	N	N	N	N	0
2	Atkins and Jones (1999)	M	N	M	N	S	N	N	N	N	4
3	Atkins and Jones (2008)	S	N	M	N	S	N	N	N	N	5
4	Bailar, Kleinberg, Castellion, Moeller, Guss and Metz (1984)	N	N	N	N	N	N	N	N	N	0
5	Bishop (2002)	N	N	N	N	N	N	N	N	N	0
6	Bodner and Pardue (1989)	N	N	N	N	N	N	N	N	N	0
7	Boikess and Edelson (1985)	S	N	S	N	M	N	N	N	M	6
8	Brady (2000)	M	N	M	N	N	N	N	N	N	2
9	Brady and Holum (1981)	M	N	M	N	M	N	N	N	N	3
10	Brady and Holum (1996)	M	N	M	N	M	N	N	N	N	3
11	Brady and Humiston (1996)	M	N	N	N	M	N	N	N	N	2
12	Brady, Russell and Holum (2000)	M	N	M	N	S	N	N	N	N	4
13	Brescia, Arents, Meislich and Turk (1975)	N	N	M	M	N	N	N	N	N	2
14	Brown and LeMay (1985)	N	N	M	N	N	M	M	N	S	5
15	Brown, LeMay, Bursten and Burdge (2003)	N	N	N	N	N	N	N	N	N	0
16	Burns (1995)	M	N	N	N	N	N	N	N	N	1
17	Chang (1994)	N	N	M	N	N	N	N	N	N	1
18	Chang (2003)	N	N	M	N	N	N	N	N	N	1
19	Daub and Seese (1996)	M	N	N	N	N	N	N	N	N	1
20	Dickerson, Gray and Haight (1974)	N	N	M	N	N	N	N	N	N	1
21	Dickerson, Gray, Darensbourg and Darensbourg (1984)	M	N	N	N	N	S	N	N	N	3
22	Dickson (2000)	S	N	N	N	S	N	N	N	S	6
23	Ebbing and Wrighton (1993)	S	N	M	N	M	N	N	N	N	4
24	Garland (1975)	M	N	N	N	M	M	N	N	N	3
25	Gillespie, Humphries, Baird and Robinson (1989)	N	N	N	N	N	N	N	N	N	0
26	Goates, Ott and Butler (1981)	S	M	M	N	N	N	N	N	N	4
27	Goldberg (2001)	N	N	N	N	N	N	N	N	N	0
28	Gray and Haight (1969)	M	S	M	N	M	N	N	N	N	5
29	Hein (1990)	M	N	M	N	N	N	N	N	N	2
30	Henold and Walmsley (1984)	N	N	M	N	N	N	N	N	N	1
31	Hill (1975)	S	N	M	N	M	N	N	N	N	4
32	Hill and Feigl (1987)	M	N	M	N	M	N	N	N	N	3
33	Hill and Petrucci(1999)	M	N	M	N	S	N	N	N	N	4
34	Holum (2000)	N	N	M	N	N	N	N	N	N	1
35	Joesten, Johnston, Netterville and Wood (1991)	M	N	S	N	S	N	N	N	N	5
36	Jones and Atkins (2000)	N	N	N	N	S	N	N	N	N	2
37	Keenan, Kleinfelter and Wood (1985)	N	N	N	N	N	N	N	N	N	0
38	Kotz and Treichel (1999)	M	N	M	N	S	N	N	N	S	6
39	Lippincott, Garrett and Verhoek (1977)	N	N	N	N	N	N	N	N	N	0
40	Mahan and Myers (1990)	N	N	N	N	N	N	N	N	N	0
41	Malone (2001)	S	N	N	N	N	N	N	N	N	2
42	Masterton and Slowinski (1980)	N	N	N	N	N	N	N	N	N	0
43	Masterton, Slowinski and Stanitski (1985)	N	N	N	N	N	N	N	N	N	0
44	Matta and Wilbraham (1981)	N	N	N	N	N	N	N	N	N	0
45	McMurry and Fay (1998)	M	M	N	N	N	N	N	N	N	2

(continued)

Table 1.1 (continued)

No.	Textbook	Criteria[a]									Points[b]
		1	2	3	4	5	6	7	8	9	
46	Miller (1984)	M	N	N	N	M	N	S	S	N	6
47	Moore, Davies and Collins (1978)	N	N	N	N	N	N	N	N	N	0
48	Moore, Stanitski and Jurs (2002)	M	N	N	N	N	N	N	N	N	1
49	Mortimer (1983)	M	N	N	N	N	N	N	S	N	3
50	Murphy and Rousseau (1980)	N	N	N	N	M	N	N	N	N	1
51	Ouellette (1975)	S	N	N	N	N	N	N	N	N	2
52	Oxtoby, Gillis and Nachtrieb (1999)	N	N	N	N	N	N	N	N	N	0
53	Oxtoby, Nachtrieb and Freeman (1990)	N	N	N	N	N	N	N	N	N	0
54	Peters (1990)	N	N	M	N	S	N	N	N	N	3
55	Petrucci (1989)	S	N	M	N	S	S	N	N	N	7
56	Petrucci, Harwood and Herring (2003)	M	N	S	N	S	N	M	N	N	6
57	Phillips, Strozak and Wistrom (2000)	M	N	N	N	N	N	N	N	N	1
58	Quagliano and Vallarino (1969)	M	N	N	M	N	N	N	N	N	2
59	Russo and Silver (2002)	M	N	N	M	N	N	N	N	N	2
60	Segal (1989)	N	N	N	N	N	N	N	N	N	0
61	Sherman, Sherman and Russikoff (1992)	N	N	N	N	N	N	N	N	N	0
62	Silberberg (2000)	N	N	M	N	M	N	N	S	N	4
63	Sisler, Dresdner and Mooney (1980)	N	N	M	N	N	N	N	N	N	1
64	Slabaugh and Parsons (1976)	M	N	M	N	N	N	M	N	N	3
65	Spencer, Bodner and Rickard (1999)	N	N	N	N	N	N	N	N	N	0
66	Stoker (1990)	M	N	M	N	N	N	N	N	N	2
67	Summerlin (1981)	S	N	N	N	N	N	N	N	N	2
68	Toon and Ellis (1978)	S	N	N	S	S	S	S	N	N	10
69	Tro (2008)	S	M	S	N	M	N	N	M	N	7
70	Ucko (1982)	M	N	N	N	M	M	N	N	N	3
71	Umland and Bellama (1999)	N	N	N	N	N	N	S	N	N	2
72	Whitten, Davis and Peck (1996)	M	N	N	N	M	N	M	N	N	3
73	Williams, Embree and DeBey (1981)	M	N	N	N	N	N	M	N	N	2
74	Wolfe (1988)	N	N	N	N	N	N	S	N	S	4
75	Zumdahl (1993)	S	N	M	M	M	N	S	N	S	9

[a] Criteria
1. Tentative Nature of Scientific Theories
2. Laws and Theories Serve Different Roles in Science (theories do not become laws even with additional evidence)
3. There is no Universal Step-by-Step Scientific Method
4. Observations are Theory-Laden
5. Scientific Knowledge Relies Heavily, but not Entirely, on Observation, Experimental Evidence, Rational Arguments, Creativity and Skepticism
6. Scientific Progress is Characterized by Competition between Rival Theories
7. Scientists can Interpret the same Experimental Data Differently
8. Development of Scientific Theories, at times is based on Inconsistent Foundations
9. Scientific Ideas are Affected by their Social and Historic Milieu
S satisfactory, M mention, N no mention
[b] Points: $S = 2$, $M = 1$, $N = 0$

Table 1.2 Distribution of general chemistry textbooks according to criteria and classification ($n = 75$)

Criteria	Classification[a]		
	N (%)	M (%)	S (%)
1	33 (44.0)	29 (38.7)	13 (17.3)
2	71 (94.7)	3 (4.0)	1 (1.3)
3	44 (58.7)	27 (36.0)	4 (5.3)
4	69 (92.0)	5 (6.7)	1 (1.3)
5	47 (62.7)	16 (21.3)	12 (16.0)
6	70 (93.3)	3 (4.0)	2 (2.7)
7	64 (85.3)	5 (6.7)	6 (8.0)
8	71 (94.7)	1 (1.3)	3 (4.0)
9	69 (92.0)	1 (1.3)	5 (6.7)

[a] Classification: S satisfactory, M mention, N no mention

Consider the various models that have been used by scientists as aids to their under-standing of the atom. An early model, [Dalton] represented the atom as a hard, incom-prehensible sphere. The model fits many observations. The hard-sphere model does not fit certain other observations, however. [At this stage the authors present the atomic models of Thomson, Rutherford, Bohr, and wave-mechanical]. The history of the atomic model shows clearly the kind of changes that occur with a model in science. Each successive model is capable of demonstrating that which was shown by its predecessor, and in addition it comes closer to giving a complete demonstration of the behavior of nature (Goates et al. 1981, pp. 3–4, italics in original).

Experimental data and observations also lead to the development of chemical concepts, theories and models which help us to understand our observations. For example, the first modern concept of an atom and the atomic theory were developed by John Dalton. Atoms are far too small to be observed directly. The best we can do is to develop a tentative mental picture of the concept. These mental pictures, called *models*, help scientists to understand and explain abstract concepts. Although this model [Dalton's] is extremely useful both then and now, it is important to avoid taking models too literally. They all have limitations and fall short of reality. The model of atoms has been modified many times since Dalton's time as a result of the work and discoveries of many scientists. Let us briefly examine the evolution of the atomic model from 1800 to the present. [At this stage the authors present a brief summary of the atomic models of J.J. Thomson (1856–1940), E. Rutherford (1871–1937), N. Bohr (1885–1962), and wave-mechanical].The evolution of the atomic model from Dalton's simple 'billiard ball atoms' to the highly mathematical, abstract and sophisticated wave-mechanical model illustrates the importance of experimental investi-gation. As new evidence accumulates, theories and models must be modified accordingly. It should be noted however, that no matter how refined a model of atoms become, it can never depict a true atomic system (Toon and Ellis 1978, pp. 6–10, emphasis in original).

Observations often lead scientists to formulate a **hypothesis**, a tentative interpretation, or explanation of the observations; the scientific approach returns to observations to test theories. Theories are validated by experiments, though they can never be conclusively proved—there is always the possibility that a new observation or experiment will reveal a flaw. For example, a central theory to chemistry is John Dalton's atomic theory—the idea that all matter is composed of atoms. Is this theory 'true'? Was it reached in logical, unbiased ways? Will this theory still be around in 200 years? The answers to these questions depend on how you view science and its development. One way to view science—let us call it the *traditional view*—is the continued accumulation of knowledge and the building of increasingly precise theories . In the twentieth century, however, a different view of scientific knowledge began to develop. In particular, a book by Thomas Kuhn, entitled *The Structure of Scientific Revolutions*, challenged the traditional view (Tro 2008, pp. 5–7, emphasis and italics in original).

[The author adds that Kuhn's ideas came from his study of the history of science, which allowed him to question the traditional view and suggest instead revolutionary changes in science].

These three examples of satisfactory (S) presentations show clearly, that in order to include NOS, textbook authors can easily use the context of the development of theories in chemistry. The first two textbooks (Goates et al. 1981; Toon and Ellis 1978) explicitly refer to the atomic models of Dalton, Thomson, Rutherford, Bohr and the wave-mechanical, to illustrate how theories need to be modified continually (tentative NOS) in order to explain experimental findings. Tro (2008) in turn raises very pertinent issues, such as: (a) Is this theory 'true'? (b) Was it reached in logical, unbiased ways? and (c) Will this theory still be around in 200 years? These are novel ways to introduce chemistry to students and can arouse their curiosity and interest with respect to what is science and how it develops.

Twenty-nine textbooks were classified as Mention (M) and the following are three examples:

Some currently accepted theories will eventually be modified, and others may be replaced altogether if new experiments uncover results that present theories can't explain (McMurry and Fay 1998, p. 3).

Theory, of course, is still tentative. In the words of J.J. Thomson, the English scientist who won the 1906 Nobel prize for physics, 'From the viewpoint of the physicist, a theory is a matter of policy rather than a creed; its object is to connect or coordinate apparently diverse phenomena, and above all to suggest, stimulate, and direct experiment' (Slabaugh and Parsons 1976, p. 7).

Most theories in use have known limitations. These 'imperfect' theories are simply the best ideas anyone has found *so far* to describe, explain and predict what happens in the world in which we live. Theories with limitations are generally not abandoned until a better theory is developed (Stoker 1990, p. 6, original italics).

The main difference between textbooks that were classified as Satisfactory (S) or Mention (M), was that the latter do not refer to actual examples from the chemistry curriculum.

1.3.4 Criterion 2: Laws and Theories Serve Different Roles in Science (Theories Do Not Become Laws Even with Additional Evidence)

The objective of this criterion was to see whether textbooks presented a simplistic hierarchical relationship in which hypotheses become theories and theories become laws, depending on the amount of evidence/proof. The inverse relationship in which laws become theories would also constitute a hierarchical relationship. Results obtained (see Tables 1.1 and 1.2) show that only one of the textbooks had a Satisfactory (S) presentation, and two were classified as Mention (M). Following are the examples of textbooks that were classified as Mention (M):

> A rock dropped from the hand falls to the ground... *why* the rock falls. It is true that the magnitude of the attraction between earth and rock was expressed by Sir Isaac Newton in his law of universal gravitation, but even that well-established law explains nothing about the reasons for the attraction. It has become common for people to say that 'experiments have proved' a scientific law to be true, or that the 'law is derived from the facts.' But such statements are false. Unfortunately, the aura surrounding the terms 'scientific' and 'scientific proof' is such that it lends those phrases dignity, but they and the context in which they are often used are meaningless" (Goates et al. 1981, p. 2, italics in original).

> It is important to keep in mind as you study chemistry or any other science that scientific theories are not laws of nature. All they do is represent the best explanations of experimental results that we can come up with at the present time (McMurry and Fay 1998, p. 3).

The presentation by Goates et al. (1981) implicitly recognizes the difference between theories and laws, as the latter do not explain the reasons for gravitational attraction.

Following is an example of the only textbook that was classified as Satisfactory (S) on Criterion 2:

> It is important to understand that a law that correlates a series of observations, is essentially empirical; it only registers and summarizes in a concise manner the results of a great number of experiments ... theories explain observations according to an imaginary framework, not directly observable, and predict what has not been observed so far. For example, a law that we attribute to Boyle affirms that at low pressures the volume of a gas is inversely proportional to the pressure exerted on the vessel. A theory suggests that Boyle's law is obeyed as particles (molecules) of gaseous material are far from each other and can easily approximate in order to increase the pressure, and can draw apart if the pressure is decreased. The law is observed directly, whereas the theory must always remain as a possible explanation, until the molecules of the gases can be observed directly (Gray and Haight 1969, pp. 1–3).

The presentation by Gray and Haight (1969) comes quite close to how philosophers of science generally differentiate between theories and laws. For example, Losee (2001) differentiates between laws and theories in similar terms, and provides examples of laws of nature, such as: Boyle's law and Galileo's laws of free fall.

Following is an example of a textbook that was classified as No mention (N) on this criterion:

> Often a large number of related scientific facts can be summarized into broad, sweeping statements called **natural, or scientific, laws**. The law of gravity is a classic example of a natural law. This law—all bodies in the universe have an attraction for all other bodies that is directly proportional to the product of their masses and inversely related to the square of their separation distance—summarizes in one sweeping statement an enormous number of facts... Such a natural law can be established in our minds only by inductive reasoning; that is, you conclude that the law applies to all possible cases, since it applies in all of the cases studied or observed" (Joesten et al. 1991, p. 6, emphasis in original).

Now, let us compare this textbook presentation of Newton's law of gravitation, to that of a philosopher of science:

> For bodies which are both massive and charged, the law of universal gravitation and Coulomb's law (the law that gives the force between two charges) interact to determine the final force. But neither law by itself truly describes how the bodies behave. No charged objects will behave just as the law of universal gravitation says; and any massive objects will constitute a counterexample to Coulomb's law. *These two laws are not true: worse they are not even approximately true'* (Cartwright 1983, p. 57, emphasis added).

This leads to a dilemma: Did Newton formulate his law of gravitation based entirely on experimental observations (as most textbooks seem to suggest, inductive reasoning, Joesten et al. 1991)? If the answer is in the affirmative then Newton should have been aware that charged bodies would not follow the law of gravitation. Insight from Giere (1999) can help to resolve the dilemma:

> Most of the laws of mechanics as understood by Newton, for example, would have to be understood as containing the proviso that none of the bodies in question is carrying a net charge while moving in a magnetic field. That is not a proviso that Newton himself could possibly have formulated, but it would have to be understood as being regularly invoked by physicists working a century or more later (p. 91).

In other words, 'Newtonian method' is far from being based on inductive reasoning and is more of an idealization or abstraction of such phenomena, and does not describe the behavior of actual bodies. At this stage it is important to note that there is considerable controversy with respect to understanding laws and theories, among philosophers of science (cf. Cartwright 1983; Giere 1999; 2006; Lakatos 1970; Nagel 1961). Without going into considerable detail, some of this will be discussed in the conclusion section.

1.3.5 Criterion 3: There is No Universal Step-by-Step Scientific Method

Most science and methodology courses emphasize the importance of the scientific method and the same framework is repeated in general chemistry textbooks, in the form of flow diagrams, such as: Observations → Hypotheses → Experiments → Analysis of data → Conclusions (based on theories or laws). Some textbooks present more elaborate diagrams based on cycles. Windschitl (2004, p. 505) considers such presentations as the unproblematic scientific method. Results obtained (see Tables 1.1 and 1.2) show that only four textbooks made a Satisfactory (S) presentation and 27 a simple Mention (M). Following is an example from a textbook that was classified as No mention (N):

> Order the following terms so that they represent the normal steps of the scientific method: facts, law, theory, experiment, hypothesis (Burns 1995, problem section, p. 17) [On p. 13 the textbook had presented the flow diagram of the unproblematic scientific method].

Following are five examples of textbooks that were classified as Mention (M):

This all sounds straightforward, but in practice scientists do not sit down with a checklist to mark off each hypothetical stage of the scientific method. In fact, in many cases, theory comes before observation, and the theory can easily precede the formulation of any laws (Henold and Walmsley 1984, p. 2).

Science is not totally different from other disciplines. For example, creativity is central to both science and the humanities. Science does not simply involve cold logic to the exclusion of more human characteristics. Albert Einstein recognized that there was no *logical* path to some of the laws that he formulated. Even he relied on intuition, based on experience and understanding. There is, then, no single 'scientific method' which when followed, produces guaranteed results. Scientists observe, gather facts, make hypotheses, but somewhere along the way they test their hunches and their organization of facts by experimenting. Scientists, like other human beings, use intuition and generalize from few facts. Sometimes they are wrong (Hill 1975, pp. 5–6, original italics). [This textbook was not classified as satisfactory as although it recognized that there is no single scientific method, it implied that there could be various such methods. Furthermore, it does not explicitly provide examples].

There is no single scientific method that serves as a guideline for all to follow. Like everyone, scientists make missteps and sometimes great leaps of the imagination. Often, though, they do proceed rather methodically, checking their ideas with carefully designed experiments (Hill and Petrucci 1999, p. 8).

... to provide a small glimpse of the character of chemistry. The observing, hypothesizing, testing, and retesting, theorizing, and finally, reaching conclusions have been going on for centuries. And they continue today more actively than ever before. Collectively they are often called the *scientific method*. There is really no rigid order to the scientific method. Looking back over the history of science, though, the preceding features always seem to be present (Peters 1990, p. 3, original italics).

... One of the foremost American physicists, P.W. Bridgman, once said that the scientific method is nothing more than doing one's level best with one's mind, no holds barred. James B. Conant, a noted chemist and former president of Harvard University, defined it as the 'tactics and strategy of science.' What most distinguishes scientific thinking is the mental attitude of adventuring investigators, who pursue their intellectual quests with intense curiosity, utmost thoroughness, and a healthy skepticism of their own results (Sisler et al. 1980, p. 5).

Textbooks that were classified as Mention (M) clearly went beyond the traditional recipe-like scientific method, by emphasizing intuition, going beyond the cold logic, leaps of imagination, intellectual quests, curiosity and healthy skepticism. History of science shows that these facets have indeed played an important role in scientific progress (Niaz 2009). Furthermore, these textbooks endorsed their understanding of the scientific method by referring to Einstein, Bridgman, and Conant. The role of Conant (1945, 1947) is particularly welcome in recognizing the importance of history of science for science education and also for his influence on Thomas Kuhn.

Following are three examples of textbooks that were classified as Satisfactory (S):

One last word about the scientific method: some people wrongly imagine science to be a strict set of rules and procedures that automatically lead to inarguable, objective facts. This is not the case. Even our diagram [quite similar to that presented in most textbooks] of the scientific method is only an idealization of real science, useful to help us see the key distinctions of science. Doing real science requires hard work, care, creativity, and even a bit of luck. Scientific theories do not just fall out of data—they are crafted by men and women of great genius and creativity. A great theory is not unlike a master painting and many see a similar kind of beauty in both. (For more on this aspect of science, see the box entitled *Thomas S. Kuhn and Scientific Revolutions*). (Tro 2008, p. 6, italics in original)

It is not correct to suppose that success in science is guaranteed by simply following a series of procedures similar to a recipe book. Sometimes, scientists develop procedures for reasoning in their area of work, known as a *paradigm*, whose success is great at the beginning, but later it is not so. Eventually, a new paradigm becomes necessary ... Lastly, many discoveries are accidental (X-rays, radioactivity, penicillin, to mention a few) ... Perhaps, nobody has been more conscious of this than Louis Pasteur, who wrote: 'Chance favors the prepared mind' (Petrucci et al. 2003, pp. 3–4, original italics).

The third example comes from a textbook that had a similar presentation (classified as S) and significantly added the following comment about accidental discoveries:

The use of the term *serendipity* for accidental discoveries was first proposed in 1754 by Horace Walpole after he read a fairy tale titled 'The Three Princes of Serendip.' Serendip was the ancient name of Ceylon and the princes, according to Walpole, 'were always making discoveries by accident, of things they were not in quest of'(Joesten et al. 1991, p. 8, original italics).

These textbooks explicitly argued against a rigid scientific method leading to objective facts. Tro (2008) for example provides the traditional diagram of the scientific method and then points out that it is an idealization of real science. Furthermore, the author emphasized that theories do not 'fall out of data' but on the contrary require a considerable amount of hard work and creativity. Of all the textbooks analyzed in this study, Tro (2008) was the only one to not only cite Kuhn (1962), but also discuss some of its implications, in a section entitled: "The Nature of Science: Thomas S. Kuhn and Scientific Revolutions". This should be a cause of concern to science educators, as Kuhn's major work was published almost 45 years ago. Of course, reading Kuhn does not mean that one has to agree with him.

1.3.6 Criterion 4: Observations are Theory-Laden

Results obtained (see Tables 1.1 and 1.2) show that only one textbook had a Satisfactory (S) presentation, five Mentioned (M) and 69 textbooks made No mention (N) of this criterion. Following are two examples of textbooks that were classified as No mention (N):

The true scientist does not gather facts to support his preconceived theories. This is, in fact, the crucial difference between the early philosophers and the more modern scientist (Summerlin 1981, p. 4).

A scientist must be ready to remove himself from his previous training, ideas and personal prejudices so that faulty models are not promulgated beyond their usefulness (Ouellette 1975, p. 119).

Following are examples of two textbooks that were classified as Mention (M):

Scientists are aware that beyond the limits of their ability to observe the physical world, they undergo significant subjective experiences to which objective scientific methods cannot be adequately applied and of which scientific facts and theories are completely independent (Quagliano and Vallarino 1969, p. 2).

The coupling of observations and hypotheses occurs because once we begin to proceed down a given theoretical path, our hypotheses are unavoidably couched in the language of that theory. In other words, we tend to see what we expect to see and often fail to notice things that we do not expect. Thus the theory we are testing helps us because it focuses our questions. However, at the very same time, this focusing process may limit our ability to see other possible explanations (Zumdahl 1993, pp. 5–7).

Following is an example of the only textbook that was classified as Satisfactory (S):

Test your awareness: First read the sentence enclosed in the box below

> FINISHED FILES ARE THE RESULT OF YEARS OF SCIENTIFIC STUDY COMBINED WITH THE EXPERIENCE OF MANY YEARS

Now count the F's in the sentence. Count them only once and do not go back and count them again. See discussion in margin on p. 14

Answer to alertness test. There are six F's. Because the F in 'of' sounds like a V, it seems to disappear, and most people count only three F's. It is really remarkable how frequently we fail to perceive things as they really are (Toon and Ellis 1978, p. 14).

This is an interesting way to introduce the theory-ladenness of observations, namely what we observe is influenced by our theoretical frameworks. It can be argued that a more explicit example from history of science would be more helpful for students. Interestingly, philosopher–physicist Norwood Russell Hanson (1969, p. 90) used the diagrams of duck/rabbit and young/old woman to illustrate the same idea. In the case of the young/old woman, sometimes when you look at the diagram you see the young woman and at other times the old woman. Hanson (1958) had earlier made the same point by arguing that J. Kepler (1571–1630) and T. Brahe (1546–1601), leading astronomers of their time may have seen the same physical objects at dawn and still their conceptualizations differed.

1.3.7 Criterion 5: Scientific Knowledge Relies Heavily, but Not Entirely, on Observation, Experimental Evidence, Rational Arguments, Creativity and Skepticism

This criterion required textbooks to incorporate various aspects of the scientific endeavor ranging from observations to arguments to creativity. Results obtained (see Tables 1.1 and 1.2) show that 12 textbooks were classified as Satisfactory (S), 16 as Mention (M) and 47 as No mention (N). Following is an example of a textbook that was classified as Mention (M):

> Bacon said: *There are two methods in which we acquire knowledge – argument and experimentation. Argument allows us to draw conclusions, and may cause us to admit the conclusion, but it gives no proof, nor does it remove doubt, and cause the mind to rest in conscious possession of truth, unless the truth is discovered by means of experience...* The scientific method is a powerful tool. But it is not a guaranteed means for scientific discovery. Creativity, inventiveness, imagination, and sometimes even luck, are also needed (Ucko 1982, pp. 6–8, original italics).

Following are three examples of textbooks that were classified as Satisfactory (S):

> The tale of the discovery of DNA's structure has been told by Watson in his book *The Double Helix* ... [Rosalind] Franklin [King's College] had experimental data, but, to Watson and Crick, she seemed unwilling to share it. To compound the problem, the experiments were difficult to perform, and Watson and Crick had neither the expertise nor the equipment to do them ... it seemed to Watson and Crick that the King's College scientists did not really appreciate the significance of their work. So Watson and Crick were faced with a dilemma: How could they convince the King's College group to share their data, and more generally, was it ethical to work on a problem that others had claimed as theirs ... Watson and Crick knew from the beginning that the overall structure of DNA was a helix ... What they did not know was the detailed structure of the helix ... Creativity and insight are needed to transform a fortunate accident into useful and exciting results, wonderful examples of which are the recent discovery of the cancer drug cisplatin by Barnett Rosenberg, the discovery of penicillin by Alexander Fleming (1881–1955), or of iodine by Bernard Courtois (1777–1838). (Kotz and Treichel 1999, pp. 6–8, original italics).

> The ability to make detailed and accurate observations, coupled with intellectual curiosity, can lead to unexpected discoveries [Examples of Louis Pasteur (1822–1895) and William Perkin (1838–1907), dye industry are provided]... It is possible that you may some day make an important contribution to scientific knowledge. In your study of chemistry, you will have the opportunity to find answers by means of your own experiments. This experience should give you a great deal of personal satisfaction. In a way, you may feel the same excitement of discovery experienced by every scientist when he finds answers to his problems. You should try to develop the characteristics, attitudes, and techniques of skilful scientists, some of which are listed below: (1) An inquiring mind. (2) Accurate and critical observations. (3) Recording of data in a thorough, neat, and organized fashion. (4) Alertness to recognize the unexpected. (5) Willingness to reject old ideas and to accept new ones when sufficient data warrant it. (6) Resistance to the tendency to make generalizations on the basis of insufficient data. The greatest assets in your search for knowledge are your intellectual curiosity and your ability to reason and understand (Toon and Ellis 1978, pp. 12–13).

> Observation requires careful attention to details, whereas the development of a hypothesis requires insight, imagination and creativity. For example, although Dalton could not see individual atoms, he was able to imagine them and formulate his atomic hypothesis. It was a monumental insight that helped others understand the world in a new way (Jones and Atkins 2000, p. 6).

These examples of Satisfactory (S) presentations provide a good blueprint of how textbooks can introduce NOS based on historical episodes. Discovery of the DNA structure is an engaging and thought-provoking experience and illustrates explicitly how the scientific endeavor is a complex and a deeply human enterprise. Toon and Ellis (1978) have drawn attention for the need to make students cognizant of their own ability to make important contributions to scientific progress. Among the characteristics of scientists, the following deserve special mention: 'tendency to make generalizations on the basis of insufficient data.' Philosopher–physicist Gerald Holton has referred to this tendency with respect to the Millikan–Ehrenhaft controversy (oil drop experiment, Niaz 2005a, 2009) as 'suspension of disbelief': "... the scientist's ability during the early period of theory construction and theory confirmation to hold in abeyance final judgments concerning the validity of apparent falsifications of a promising hypothesis" (Holton 1978, p. 212). Indeed, as the Millikan–Ehrenhaft controversy showed clearly, empirical data do not unequivocally provide evidence for a particular hypothesis. Holton (1986) has lamented that science education in general does not recommend Millikan's methodology to our 'beginning students' (p. 12).

1.3.8 Criterion 6: Scientific Progress is Characterized by Competition Between Rival Theories

The role of competition between rival theories has been recognized by philosophers of science (Lakatos 1970). Results obtained (see Tables 1.1 and 1.2) show that only two textbooks were classified as Satisfactory (S), three as Mention (M), and 70 as No mention. Following are two examples of textbooks that were classified as Mention (M):

Scientists generally try to find support for their own theories and to disprove rival theories by doing experiments (Ucko 1982, p. 8).

Exchange of information does not always lead to agreement among scientists. Disagreements (sometimes quite heated) often stimulate progress. Scientists who are aware of that make no attempt to avoid arguments, but they try to keep their arguments within boundaries consistent with the scientific method (Garland 1975, p. 12).

Following is an example of a textbook that was classified as Satisfactory (S):

Petrucci (1989) provides a brief introduction as to how some scientists go beyond existing knowledge and discover the key to understanding scientific phenomena. It then refers the student to the laws of classical physics to explain black body radiation, the origin and development of the quantum theory and the roles played by Max Planck (1858–1947) and Albert Einstein. Finally, it concluded: "There have been instances in the history of science when a new hypothesis proved useful in explaining one phenomenon but was not generally applicable to any others. It was only in the discovery of other applications of the quantum hypothesis that it acquired a status as a significant new theory of science. The first notable new success came in 1905 with Albert Einstein's (1879–1955) quantum explanation of the photoelectric effect" (Petrucci 1989, p. 246).

The difference between textbooks classified as Satisfactory (S) and Mention (M) was that the former discussed the role of rival hypotheses in the context of development and origin of particular theories, which facilitates better understanding for students.

1.3.9   Criterion 7: Scientists Can Interpret the Same Experimental Data Differently

The importance of alternative interpretations of experimental data has been recognized in both the philosophy of science (Lakatos 1970) and science education (Niaz 2001c). Results obtained (see Tables 1.1 and 1.2) show that six textbooks were classified as Satisfactory (S), five as Mention (M), and 64 as No mention (N). Following is an example of textbooks classified as Metnion (M):

> When different and contradictory theories are presented, in general the one that provides better prediction is selected. The theory that requires a lesser number of presuppositions, that is a simpler theory is preferred. With the passage of time as new experimental evidence accumulates, most of the theories are modified and some discarded (Petrucci et al. 2003, p. 3).

Following are three examples of textbooks classified as Satisfactory (S):

> The importance of these activities [methods of scientists] is illustrated by comparing the work of two scientists who proposed conflicting theories about ordinary burning (combustion) (p. 11)... In the 17th century, these observations led Georg Ernst Stahl (1660–1734) to propose that all combustible materials possess a 'fire substance' called *phlogiston*. He believed that wood was composed of ash and phlogiston. When wood burned, the phlogiston escaped and the ash remained. To explain the apparent increase in mass observed when metals burned, it was necessary to say that phlogiston might sometimes have negative mass... In the 18th century, Antoine Lavoisier noted that when metals are heated in air, an apparent increase in their mass occurs... This led him to propose that combustion is a combination of a metal with a gas from the air (Toon and Ellis 1978, pp. 11–12, original italics).

> Distinguishing between observations and what you think about them are important. If observations are made carefully, you would make the same observations if you were to repeat the experiment. However, you might change your mind about what you thought about them because of something you had learned. Other people in other places or at other times would also make the same observations if they did the same experiment. But they might explain their observations differently than you. For example, scientists today usually interpret their observations in terms of atoms and molecules. Before the seventeenth century, people usually accounted for their observations in terms of religion ... the overturning of an important theory opens new frontiers in science. For example, before the sixteenth century, people thought that Earth was the center of the universe. Then Copernicus explained that observations of the planets by suggesting that Earth and the other planets revolve around the sun. This was the beginning of the scientific revolution (Umland and Bellama 1999, pp. 3–7).

Zumdahl (1993) presented a very interesting section entitled 'Observations, Theories and the Planets', along with a timeline ranging from 2000 BC to 2000 AD, and following are some of the excerpts (classified as S):

Humans have always been fascinated by the heavens, by the behavior of the sun by day and the stars by night … the basic *observations* of these events have remained the same over the past 4000 years. However, our *interpretations* of the events have changed dramatically. For example, about 2000 B.C. the Egyptians postulated that the sun was a boat inhabited by the god Ra, who daily sailed across the sky … Eudoxus, born in 400 B.C. … imagined the earth as fixed, with the planets attached to a nested set of transparent spheres that moved at different rates around the earth … Five hundred year later, Ptolemy, a Greek scholar, worked out a plan more complex than that of Eudoxus, in which the planets were attached to the edges of spheres that 'rolled around' the spheres … in 1543, a polish cleric, Nicolas Copernicus, postulated that the earth was only one of the planets, all of which revolved around the sun … Kepler postulated elliptical rather than circular orbits for the planets in order to account more completely for their observed motions. Kepler's hypotheses were in turn further refined 36 years after his death by Isaac Newton, who recognized that the concept of gravitation could account for the positions and motions of the planets … Einstein … showed that Newton's mechanics was a special case of a much more general model … (Zumdahl 1993, p. 6, original italics).

The presentation by Toon and Ellis (1978) reveals the importance of how two scientists (Stahl and Lavoisier) can interpret the same phenomenon of combustion differently. The role of the *phlogiston* theory and the chemical revolution (based on the work of Lavoisier) constitutes an important episode in the history of science and some textbooks do refer to it. However, the Toon and Ellis (1978) presentation deserves special recognition as it presents the historical episode as an instance of 'conflicting theories' and thus provides an example of how textbook authors can incorporate NOS, based on HPS. Philosophers of science consider the *Chemical Revolution* associated with *phlogiston* and Lavoisier as an important episode in the history of science and have interpreted it from various perspectives (cf. McCann 1978; Musgrave 1976; Perrin 1988; Thagard 1990). Niaz (2008b) has provided details of this controversy and discussed its educational implications for teaching chemistry (pp. 4–5, 15).

Zumdahl (1993) has provided a good overview of how our *observations* related to the heavenly bodies (a topic of interest to most students) have been *interpreted* differently for almost 4,000 years. Again, this provides a good example of how NOS aspects can be incorporated into the textbooks. This discussion leads to another important issue. If observations can have varying interpretations, based on different theories and models which lead to controversies, can we conclude that this undermines the objective NOS? To respond to this question we sought help from Leon Cooper (Nobel Laureate, physics, 1972), who responded in the following terms: "Observations can have varying interpretations, but this does not undermine the objective nature of science. It is somewhat ironic that what we like to call the meaning of a theory, its interpretation, is what changes. Think, for example, of the very different views of the world provided by quantum theory, general relativity, and Newtonian theory (Reproduced in Niaz et al. 2010b). This clearly shows the importance of alternative interpretations of observations for science education, and the presentation of Zumdahl (1993), quite similar to Leon Cooper, provides a good example.

1.3.10 Criterion 8: Development of Scientific Theories, at Times is Based on Inconsistent Foundations

It appears that the inconsistent nature of scientific theories is counterintuitive for most textbook authors and 71 (94.7%) textbooks were classified as No mention (N). Only one textbook was classified as Mention (M) and three as Satisfactory (S). Following is an example from the textbook that was classified as Mention (M):

> According to Kuhn [1962], science goes through fairly quiet periods that he calls *normal science*. In these periods, scientists make their data fit the reigning theory, or paradigm. Small inconsistencies are swept aside during periods of normal science. However, when too many inconsistencies and anomalies develop, a crisis emerges. The crisis brings about a *revolution* and a new reigning theory... Kuhn further contends that theories are held for reasons that are not always logical or unbiased, and that theories are not *true* models — in the sense of a one-to-one mapping — of the physical world" (Tro 2008, p. 7, italics in original).

This textbook could have been classified as Satisfactory (S), if it had included at least one example from the history of science to illustrate the inconsistent nature of scientific theories, and thus facilitating greater understanding.

Following is an example of a textbook classified as Satisfactory (S):

> Hooke suggested that burning substances combine with air, but unfortunately most scientists rejected this idea to embrace the **phlogiston theory**, which held sway for the next hundred years. According to this theory, combustible materials contain phlogiston, an undetectable substance that is released when the material burns... knowing that a calx [metal oxide] weighs more than the metal from which it is formed, critics asked how the *loss* of phlogiston when a metal burns could cause a *gain* in mass. Phlogistonists either dismissed the importance of weighing or proposed that phlogiston had negative mass! Some of these responses seem ridiculous to us now, but they point out that the pursuit of science, like any other human endeavor, is subject to human failings; even today it is easy to dismiss conflicting evidence than to give up an established idea... Lavoisier proposed that when a metal forms its calx, it does not lose phlogiston but rather combines with this gas, which must be a component of air (Silberberg 2000, p. 10, italics and emphasis in original, from a section entitled: The Phlogiston Fiasco and the Impact of Lavoisier).

This is indeed an interesting interpretation of the *phlogiston theory*, if we compare it to the presentation of Toon and Ellis (1978) in Criterion 7, who highlighted different interpretations of the combustion phenomenon. Silberberg (2000) on the contrary highlights the 'negative mass' of phlogiston, and its inconsistency with the increase in mass of the metal oxides after combustion. Interestingly, however, Musgrave (1976) pointed out that as early as 1630, it was common knowledge that metallic oxides weighed more than the metals from which they were prepared, and thus if, "Lavoisier's 1772 experiment refutes phlogiston theory, then phlogiston theory was *born* refuted" (p. 183, original italics). It is important to note that two textbooks (Toon and Ellis 1978; Silberberg 2000) found two different aspects of NOS in the same historical episode (phlogiston theory), and both were classified as Satisfactory (S). Another textbook in this study (Mortimer 1983) also presented a similar interpretation and was also classified as Satisfactory (S).

At this stage it would be interesting to illustrate the inconsistent nature of scientific theories from twentieth century science. Bohr's (1913) model of the atom incorporated Planck's 'quantum of action' to the classical electrodynamics of Maxwell. For many of Bohr's contemporaries and philosophers of science, this represented a contradictory 'graft' or an inconsistent foundation. According to philosopher–physicist, Margenau (1950): "… it is understandable that, in the excitement over its success, men overlooked a malformation in the theory's architecture; for Bohr's atom sat like a baroque tower upon the Gothic base of classical electrodynamics" (p. 311). Niaz (1998) has reported that of the 23 general chemistry textbooks (all published in USA) analyzed, only two described this inconsistency satisfactorily. Further details about the inconsistent nature of Bohr's famous four postulates are provided by Niaz (2011). Similarly, Lakatos (1970) has emphasized the role of inconsistencies (contradictions) in the history of science. Interestingly, Bohr's model of the atom was in turn replaced in 1916 by the Bohr–Sommerfeld model (for details see Niaz and Cardellini 2011).

1.3.11   Criterion 9: Scientific Ideas are Affected by Their Social and Historic Milieu

This criterion provided an overview of the complexities involved in the scientific enterprise, especially with respect to the interactions among scientists, peers and society. Sixty-nine textbooks were classified as No mention (N), one as Mention (M) and five as Satisfactory (S). Following is the example of the textbook classified as Mention (M):

> The progress of science is influenced by personality, money and social forces. Scientific research generally begins with a quest for a specific piece of new knowledge, chosen by a combination of social, economic, and personal influence (Boikess and Edelson 1985, pp. 3–4).

Following are two examples of textbooks classified as Satisfactory (S):

> The development of scientific theories does not always happen easily, quickly, or smoothly. Evolution of thought takes time. The modern view of the solar system, for example, took thousands of years and countless astronomical observations to develop. At times, new ideas meet significant resistance. The famous Italian scientist Galileo Galilei (1564–1642) was forced by church authorities to retract his views that Earth moved around the sun… In the early 1900s, Marie Curie, a Polish-born French scientist, was a pioneer in the newly discovered field of radioactivity. Despite her many honors, including two Nobel prizes, she was never elected to the French Academy of Sciences. Apparently she was slighted because she was Polish born and a woman. In the 1950s, Linus Pauling, an American chemist, had his passport restricted by the government and was not allowed to travel out of the United States. In the 1970s and 1980s, Andrei Sakharov, a Russian physicist, was exiled to a small Russian city and not allowed to talk with other scientists. Both Pauling and Sakharov were punished for speaking against the development of nuclear weapons… Recently, the Catholic Church admitted that Galileo was treated

Table 1.3 Comparison of general chemistry textbooks in different periods

Period	No. of textbooks	Mean points[a]
1965–1989	15	2.3
1981–1990	27	2.5
1991–2000	23	2.5
2001–2008	10	2.4
All textbooks	75	2.4

[a] All textbooks were evaluated (see Criteria section) on a scale of 0–18 points. On each criterion, textbooks were awarded the following points: Satisfactory = 2 points, Mention = 1 point, and No mention = 0 point

unfairly in the 1600s, and Marie Curie's remains were moved to an honorary grave in the Pantheon of Paris 60 years after her death" (Dickson 2000, p. 6).

… Copernicus's writings were 'corrected' by religious officials before scholars were allowed to use them … Galileo, for example, was forced to recant his astronomical observations in the face of strong religious resistance. Lavoisier, the father of modern chemistry, was beheaded because of his political affiliations. And great progress in the chemistry of nitrogen fertilizers resulted from the desire to produce explosives to fight wars. The progress of science is often affected more by the frailties of humans and their institutions than by the limitations of scientific measuring devices … (Zumdahl 1993, pp. 6–7).

Discussion of such episodes from the history of science can provide students an opportunity to glimpse the complexity of the scientific enterprise and appreciate how, "… both rationality and objectivity come in degrees and that the task of good science is to increase these degrees as far as possible" (Machamer and Wolters 2004, p. 9). In other words, the construction of knowledge requires assumptions that support reasoning within a social and cultural context (Longino 1990, p. 219).

Comparison of the evaluation of NOS related aspects, in different time periods show (see Table 1.3), that the presentation (mean points) does not improve over time. Apparently, textbook authors and publishers are either not aware of the research literature or simply not convinced of its relevance for improving science education. Interestingly, even the subsequent editions of the textbooks make no effort to incorporate NOS-related aspects (e.g., Brady and Holum 1981, 1996; Chang 1994, 2003; see Table 1.1).

1.4 Conclusions and Educational Implications

Most textbooks in this study provided little insight into the nine criteria used for evaluating the presentation of nature of science (NOS). The percentage of textbooks that were classified as No mention (N) ranged from 44 (Criterion 1) to 94.7% (Criterion 8). Despite this, some textbooks provided good examples based on history and philosophy of science (HPS), and the percentage of textbooks classified as Satisfactory (S) ranged from 1.3 (Criterion 2) to 17.3% (Criterion 1). Of the 75 textbooks analyzed in this study, the textbook by Toon and Ellis (1978) had the highest score (10 points out of the possible 18). Three other textbooks were also noteworthy in their presentations: Zumdahl (1993), nine points; Petrucci (1989) and Tro (2008), seven points each. These examples show that although presentation of NOS is not the major objective of general chemistry textbooks, some of them inevitably refer to the historical record that aligns with the reform documents, endorsed by the American Association for the Advancement of Science (AAAS 1993) and thus provide guidelines for future textbooks. Most textbooks in this study endorsed the unproblematic traditional scientific method, the subject of Criterion 3 (58.7% were classified as N). Jenkins (2007) traced the origin of the scientific method as a political construct in the nineteenth century, which is at odds with developments in history and philosophy of science in the twentieth century. More recently, Windschitl et al. (2008) suggested going beyond the scientific method as it continues to reinforce a kind of cultural lore about what it means to participate in inquiry and lacks epistemic framing relevant to the discipline. The importance of understanding how the same observations can be interpreted differently (Criterion 7) has been illustrated very cogently by Zumdahl (1993) in the case of heavenly bodies, "Thus the same observations were made for several thousand years, but the explanations—the models—have changed remarkably from the Egyptians' boat of Ra to Einstein's relativity" (p. 6). Interestingly, Leon Cooper (Nobel Laureate in physics) also endorsed a similar vision of NOS (cf. Niaz et al. 2010b).

1.4.1   History of Chemistry is 'Inside' Chemistry

Research in science education has generally espoused the inclusion of history of science in the curriculum. Bevilacqua and Bordoni (1998) on the contrary have argued that, "We are not interested in adding the history of physics to teaching physics, as an optional subject: the history of physics is 'inside' physics" (p. 451). Similarly, Niaz and Rodríguez (2001) endorsed that history and philosophy of science are already 'inside' chemistry. Satisfactory (S) presentation of general chemistry textbooks on all nine criteria shows that this is indeed the case and following are some of the examples:

(a)   Criterion 1: Tentative nature of scientific theories is illustrated by two textbooks (Goates et al. 1981; Toon and Ellis 1978), by going into considerable detail to present the atomic models of Dalton, Thomson, Rutherford, Bohr and wave-mechanical.
(b)   Criterion 2: Role of laws and theories is illustrated by referring to how Boyle's law can be observed, whereas kinetic theory is a possible explanation (Gray and Haight 1969).
(c)   Criterion 3: Scientific method is only an idealization of real science (Joesten et al. 1991; Petrucci et al. 2003; Tro 2008).
(d)   Criterion 4: Theory-ladenness of observations, namely what we observe is influenced by our theoretical frameworks (Toon and Ellis 1978).

(e)   Criterion 5: Role of insight, imagination, and creativity in science, e.g., discovery of DNA (Jones and Atkins 2000; Kotz and Treichel 1999; Toon and Ellis 1978).

(f)   Criterion 6: Role of rival theories, based on origin and development of the quantum theory (Petrucci 1989).

(g)   Criterion 7: Alternative interpretations of experimental data, e.g., understanding heavenly bodies for 4,000 years; Phlogiston theory and chemical revolution (Toon and Ellis 1978; Umland and Bellama 1999; Zumdahl 1993).

(h)   Criterion 8: Inconsistent foundations of scientific theories, e.g., phlogiston theory was accepted despite the fact that metal oxides weigh more than metals (Silberberg 2000).

(i)   Criterion 9: Social and historic milieu, e.g., Galileo, Copernicus and the church; Pauling, Sakharov and nuclear weapons (Dickson 2000; Zumdahl 1993).

The most important feature of these examples is that these textbooks do so in a particular context of NOS. For example textbook authors in the introductory chapter have referred to a wide range of topics, such as: discovery of DNA; quantum theory; understanding heavenly bodies for 4,000 years; phlogiston theory and chemical revolution; Galileo, Copernicus and the Church; Pauling, Sakharov and nuclear weapons. By all means this is an ambitious agenda and a clear illustration of how history of chemistry can facilitate a better understanding of NOS.

In order to provide a better understanding of the role played by the introductory chapter (in textbooks), Table 1.4 presents a comparison of textbook evaluations based on three different studies. For example, in this study based on evaluation of NOS, seven textbooks had a Satisfactory (S) presentation on at least one criterion. One textbook (Dickson 2000) had Satisfactory (S) presentations on three different criteria. Brito et al. (2005) evaluated the periodic table in general chemistry textbooks, and all textbooks had a Satisfactory (S) presentation on at least one criterion. Two textbooks (Bodner and Pardue 1989; Brady and Holum 1981) had Satisfactory (S) presentations on three different criteria. Niaz and Fernández (2008) evaluated quantum numbers in general chemistry textbooks, and only two textbooks (Spencer et al. 1999; Umland and Bellama 1999), had a Satisfactory (S) presentation. This comparison shows that the profile of textbook evaluations in three different topics is quite different, that is an author may not present a consistent NOS perspective in all chapters. Consequently, evaluation of the introductory chapter provides additional information that may not be available in other chapters of the textbook.

1.4.2   Scientific Laws as Idealizations

Scientific laws being epistemological constructions do not describe the behavior of actual bodies. Galileo's law of free fall, Newton's laws, gas laws—all describe the behavior of ideal bodies that are abstractions from the evidence of experience.

Table 1.4 Comparison of general chemistry textbook evaluations based on the number of satisfactory presentations on different topics and criteria

No.	Textbook	This study	Brito et al. (2005)	Niaz and Fernández (2008)
1	Bodner and Pardue (1989)	None	3	None
2	Brady (2000)	None	1	None
3	Brady and Holum (1981)	None	3	None
4	Brady and Humiston (1996)	None	2	None
5	Burns (1995)	None	1	None
6	Daub and Seese (1996)	None	1	None
7	Dickerson et al. (1984)	1	2	None
8	Dickson (2000)	3	1	None
9	Goldberg (2001)	None	2	None
10	Hill and Petrucci (1999)	1	2	None
11	Jones and Atkins (2000)	1	2	None
12	Lippincott et al. (1977)	None	2	None
13	Malone (2001)	1	2	None
14	Masterton et al. (1985)	None	2	None
15	McMurry and Fay (2001)	None	2	None
16	Moore et al. (2002)	None	1	None
17	Phillips et al. (2000)	None	2	None
18	Quagliano et al. (1969)	None	2	None
19	Russo and Silver (2002)	None	2	None
20	Segal (1989)	None	1	None
21	Silberberg (2000)	1	2	None
22	Sisler et al. (1980)	None	1	None
23	Spencer et al. (1999)	None	1	1
24	Stoker (1990)	None	1	None
25	Umland and Bellama (1999)	1	2	1
26	Whitten et al. (1996)	None	2	None

Notes
1. This study deals with nature of science (NOS)
2. Brito et al. (2005) deals with the topic of periodic table
3. Niaz and Fernández (2008) deals with the topic of quantum numbers
4. This is a selected set of textbooks that were evaluated in all three studies

Most philosophers of science would perhaps agree that scientific laws describe the observables, whereas theories provide an explanation. In this study only one textbook provided a Satisfactory (S) presentation (Gray and Haight 1969) on Criterion 2. At this stage it would be interesting to go beyond and consider an alternative, based on Lakatos (1970) in which the clash is not between theories and facts but between two types of theories: (a) an *interpretative theory* to provide the facts and (b) an *explanatory theory* to explain them. Giere (1999) has expressed a similar thesis: "… understood as general claims about the world, most purported laws of nature are in fact false [also see Cartwright 1983, 1999]. So we need a portrait of science that captures our everyday understanding of success without invoking laws of nature understood as true, universal generalizations" (p. 24). In other words, scientific progress is characterized by a series of theories or models (plausible explanations), which vary in the degree to which they explain/interpret/ predict the experimental findings.

1.4.3 Chemistry: A Quantitative Science?

Toon and Ellis (1978, p. 14) included a section entitled: 'Chemistry is a quantitative science' with the following quote from Lord Kelvin (a leading nineteenth century British physicist):

When you can measure what you are speaking about and express it in numbers, you know something about it, and when you cannot measure it, when you cannot express it in numbers, your knowledge is of a meager and unsatisfactory kind (original italics).

Indeed, empirical data (quantitative imperative) play an important role in chemistry and most textbooks follow this line of reasoning. However, it would be too simplistic to suggest to students that chemistry is a quantitative science, with no further explanation. Holton (1978) has reasoned cogently with respect to this dilemma: "… the graveyard of science is littered with those who did not suspend belief while the data were pouring in" (p. 212). In other words, there will always be a confrontation between the quantitative imperative and the imperative of presuppositions in order to understand the empirical data (for details see, Niaz 2001c 2005b).

1.4.4 Characteristics of Scientists

Toon and Ellis (1978, p. 13) was the only textbook in this study that explicitly reminded students that someday they too could become scientists and thus feel the thrill and excitement of discovering new things. As such they made a series of six recommendations based on characteristics of working scientists, of which the following is the most significant: Resistance to the tendency to make generalizations on the basis of insufficient data. Another textbook (Tro 2008) explained that according to Kuhn a new theory is seldom or never just an increment of what is already known, and then asked students to consider the following question: "Can you think of instances in which a new theory or model was drastically different from the one it replaced?" (p. 7). Such questions can easily arouse students' curiosity, ability to reason, understand and foster interest in NOS.

1.4.5 Are We Teaching Science as Practiced by Scientists?

Schwartz and Lederman (2008) report that scientists in their study acknowledged the influence of the current scientific theory and paradigm in directing scientific research and that observations are theory-laden. Laudan (1996) argued for the need to include the history of science and its practice, "... the budding chemist learns Prout's and Avogadro's hypotheses, and Dalton's work on proportional combinations; he learns how to do Millikan's oil drop experiment; he works through Linus Pauling's struggles with the chemical bond" (p. 495). Similarly, Holton (2003) endorsed a similar agenda, "... those science educators who can be persuaded to turn to the history and philosophy of science can find fascinating material with which to infuse their own activity" (p. 604). A review of the literature, however, shows that most textbook authors, curriculum developers and even some scientists ignore the historical record and do not teach science as practiced by scientists (Niaz 2010). Nevertheless, it is important to point out that such a teaching strategy will have to be based on NOS, within the HPS perspective.

1.4.6 Author or Publisher Effect

The AAAS (1993) has advocated changing the mindset of both authors and publishers, in order to align it with the science education benchmarks (Stern and Roseman 2004). Textbook publishers form part of the 'big business' and changing their marketing practices (publisher effect) is perhaps not so straightforward. For example, Holton (2003) is not very optimistic on this account: "Most textbook publishers, who in the United States are effectively acting as the Ministry of Education, are very unlikely to allow space in a science text for more than historical anecdotes" (p. 604). The other alternative (author effect) seems to be more feasible by convincing textbook authors to pay more attention to research in science education, especially that dealing with HPS-based textbook analyses. Abd-El-Khalick et al. (2008) suggested that improving textbooks entails considerable amount of work at different levels and as such both authors and publishers need to be involved.

References

Abd-El-Khalick, F. (2005). Developing deeper understandings of nature of science: The impact of a philosophy of science course on preservice science teachers' views and instructional planning. *International Journal of Science Education, 27*, 15–42.
Abd-El-Khalick, F., Waters, M., & Le, A. (2008). Representations of nature of science in high school chemistry textbooks over the past four decades. *Journal of Research in Science Teaching, 45*, 835–855.
Akerson, V. L., Buzzelli, C. A., & Donnelly, L. A. (2010). On the nature of teaching nature of science: Preservice early childhood teachers' instruction in preschool and elementary settings. *Journal of Research in Science Teaching, 47*, 213–233.
American Association for the Advancement of Science, AAAS. (1993). *Benchmarks for science literacy, Project 2061*. New York: Oxford University Press.

Arriassecq, I., & Greca, I. M. (2007). Approaches to the teaching of special relativity theory in high school and university textbooks of Argentina. *Science and Education, 16*, 65–86.

Bauer, H. H. (1994). *Scientific literacy and the myth of the scientific method*. Champaign: University of Illinois Press.

Bevilacqua, F., & Bordoni, S. (1998). New contents for new media: Pavia project physics. *Science and Education, 7*, 451–469.

Blanco, R., & Niaz, M. (1997). Epistemological beliefs of students and teachers about the nature of science: From 'baconian inductive ascent' to the 'irrelevance' of scientific laws. *Instructional Science, 25*, 203–231.

Bodner, G. M., & Pardue, H. L. (1989). *Chemistry an experimental science*. New York: Wiley.

Bohr, N. (1913). On the constitution of atoms and molecules. *Philosophical Magazine, 26*, 1–25.

Boikess, R., & Edelson, E. (1985). *Chemical principles* (3rd ed.). New York: Harper and Row.

Brady, J. E., & Holum, J. R. (1981). *Fundamentals of chemistry*. New York: Wiley.

Brady, J. E., & Holum, J. R. (1996). *Chemistry: The study of matter and its changes* (2nd ed.). New York: Wiley.

Brito, A., Rodríguez, M. A., & Niaz, M. (2005). A reconstruction of development of the periodic table based on history and philosophy of science and its implications for general chemistry textbooks. *Journal of Research in Science Teaching, 42*, 84–111.

Burbules, N. C., & Linn, M. C. (1991). Science education and philosophy of science: Congruence or Contradiction? *International Journal of Science Education, 13*, 227–241.

Burns, R. A. (1995). *Fundamentals of chemistry (2nd ed., Spanish)*. Englewood Cliffs: Prentice Hall.

Cartwright, N. (1983). *How the laws of physics lie*. Oxford: Clarendon Press.

Cartwright, N. (1999). *The dappled world: A study of the boundaries of science*. Cambridge: Cambridge University Press.

Chang, R. (1994). *Chemistry* (4th ed., Spanish ed.). New York: McGraw Hill.

Chang, R. (2003). *Chemistry* (7th ed.,Spanish ed.). New York: McGraw Hill.

Chiappetta, E. L., & Fillman, D. A. (2007). Analysis of five high school biology textbooks used in the United State for inclusion of nature of science. *International Journal of Science Education, 29*, 1847–1868.

Chiappetta, E. L., Ganesh, T. G., Lee, Y. H., & Phillips, M. C. (2006). Examination of science textbook analysis research conducted on textbooks published over the last 100 years in the United States. Paper presented at the annual conference of the National Association for Research in Science Teaching (NARST), San Francisco, April.

Chiappetta, E. L., Sethna, G. H., & Fillman, D. A. (1991). A quantitative analysis of high school chemistry textbooks for scientific literacy themes and expository learning aids. *Journal of Research in Science Teaching, 28*, 939–951.

Cobern, W. W., Gibson, A. T., & Underwood, S. A. (1999). Conceptualizations of nature: An interpretive study of 16 ninth graders' everyday thinking. *Journal of Research in Science Teaching, 36*, 541–564.

Conant, J. B. (1945). *General education in a free society: Report of the Harvard Committee*. Cambridge: Harvard University Press.

Conant, J. B. (1947). *On understanding science*. New Haven: Yale University Press.

Crowther, J. G. (1910). On the scattering of homogeneous β-rays and the number of electrons in the atom. *Proceedings of the Royal Society, London, lxxxiv*, 226–247.

Dagher, Z. R., & BouJaoude, S. (1997). Scientific views and religious beliefs of college students: The case of biological evolution. *Journal of Research in Science Teaching, 34*, 429–445.

De Berg, K. C. (2006). The kinetic-molecular and thermodynamic approaches to osmotic pressure: A study of dispute in physical chemistry and the implications for chemistry education. *Science and Education, 15*, 495–519.

De Berg, K. C. (2008). The concepts of heat and temperature: The problem of determining the content for the construction of an historical case study which is sensitive to nature of science issues and teaching-learning issues. *Science and Education, 17*, 75–114.

Dickson, T. R. (2000). *Introduction to chemistry* (8th ed.). New York: Wiley.

Dogan, N., & Abd-El-Khalick, F. (2008). Turkish grade 10 students' and science teachers' conceptions of nature of science: A national study. *Journal of Research in Science Teaching, 45*(10), 1083–1112.

Feyerabend, P. (1993). *Against method*. New York: Verso.

Garland, J. K. (1975). *Chemistry of our world*. New York: Macmillan.

Geiger, H., & Marsden, E. (1909). On a diffuse reflection of the alpha particles. *Proceedings of the Royal Society, London, lxxxii*, 495–500.

Giere, R. N. (1999). *Science without laws*. Chicago: University of Chicago Press.

Giere, R. N. (2006). *Scientific perspectivism*. Chicago: University of Chicago Press.

Gillespie, R. J., Humphreys, D. A., Baird, N. C., & Robinson, E. A. (1989). *Chemistry* (2nd ed.). Boston: Allyn and Bacon.

Goates, J. R., Ott, J. B., & Butler, E. A. (1981). *General chemistry: Theory and description*. New York: Harcourt Brace Jovanovich.

Gray, H. B., & Haight, G. P. (1969). *Basic principles of chemistry* (Spanish ed.). New York: Benjamin.

Guisasola, J., Almudí, J. M., & Furió, C. (2005). The nature of science and its implications for physics textbooks: The case of classical magnetic field theory. *Science and Education, 14*, 321–338.

Hanson, N. R. (1958). *Patterns of discovery*. Cambridge: Cambridge University Press.

Hanson, N. R. (1969). *Perception and discovery*. San Francisco: Freeman, Cooper and Co.

Henold, K. L., & Walmsley, F. (1984). *Chemical principles, properties, and reactions*. Reading: Addison-Wesley.

Hill, J. W. (1975). *Chemistry for changing times* (2nd ed.). Minneapolis: Burgess Publishing Company.

Hill, J. W., & Petrucci, R. H. (1999). *General chemistry: An integrated approach* (2nd ed.). Upper Saddle River: Prentice Hall.

Hodson, D. (2009). *Teaching and learning about science: Language theories, methods, history, traditions, and values*. Rotterdam: Sense Publishers.

Holton, G. (1978). Subelectrons, presuppositions and the Millikan–Ehrenhaft dispute. *Historical Studies in the Physical Sciences, 9*, 161–224.

Holton, G. (1986). *The advancement of science and its burdens*. Cambridge: Cambridge University Press.

Holton, G. (2003). What historians of science and science educators can do for one another. *Science and Education, 12*, 603–616.

Irez, S. (2006). Are we prepared? An assessment of preservice science teacher educators' beliefs about nature of science. *Science Education, 90*, 1113–1143.

Irez, S. (2009). Nature of science as depicted in Turkish biology textbooks. *Science Education, 93*, 422–447.

Jenkins, E. (2007). School science: A questionable construct? *Journal of Curriculum Studies, 39*(3), 265–282.

Joesten, M, D., Johnston, D. O., Netterville, J. T., & Wood, J. L. (1991).*World of chemistry*. Philadelphia: Saunders

Jones, L., & Atkins, P. (2000). *Chemistry: Molecules, matter and changes* (4th ed.). New York: Freeman.

Justi, R., & Gilbert, J. (2000). History and philosophy of science through models: Some challenges in the case of the "atom." *International Journal of Science Education, 22*, 993–1009.

Kang, S., Scharmann, L. C., & Noh, T. (2005). Examining students' views on the nature of science: Results from Korean 6th, 8th, and 10th graders. *Science and Education, 89*, 314–334.

Khishfe, R. (2008). The development of seventh graders' views of nature of science. *Journal of Research in Science Teaching, 45*, 470–496.

Koliopoulos, D., & Constantinou, C. (2005). The pendulum as presented in school science textbooks of Greece and Cyprus. *Science and Education, 14*, 59–73.

Kotz, J. C., & Treichel, P. (1999). *Chemistry and chemical reactivity* (4th ed.). Philadelphia: Saunders.

Koul, R., & Dana, T. M. (1997). Contextualized science for teaching science and technology. *Interchange, 28,* 121–144.

Kuhn, T. S. (1962). *The structure of scientific revolutions.* Chicago: University of Chicago Press.

Lakatos, I. (1970). Falsification and the methodology of scientific research programmes. In I. Lakatos and A. Musgrave (Eds.), *Criticism and the growth of knowledge* (pp. 91–195). Cambridge: Cambridge University Press.

Laudan, L. (1996). *Beyond positivism and relativism.* Boulder: Westview Press.

Laudan, R., Laudan, L., & Donovan, A. (1988). Testing theories of scientific change. In A. Donovan, L. Laudan, and R. Laudan (Eds.), *Scrutinizing science: Empirical studies of scientific change* (pp. 3–44). Dordrecht: Kluwer.

Lederman, N. G. (1992). Students' and teachers' conceptions of the nature of science: A review of the research. *Journal of Research in Science Teaching, 29,* 331–359.

Lederman, N. G., Abd-El-Khalick, F., Bell, R. L., & Schwartz, R. (2002). Views of nature of science questionnaire: Toward valid and meaningful assessment of learners' conceptions of nature of science. *Journal of Research in Science Teaching, 39,* 497–521.

Leite, L. (2002). History of science in science education: Development and validation of a checklist for analyzing the historical content of science textbooks. *Science and Education, 11,* 333–359.

Longino, H. E. (1990). *Science as social knowledge: Values and objectivity in scientific inquiry.* Princeton: Princeton University Press.

Longino, H. E. (2004). How values can be good for science. In P. Machamer and G. Wolters (Eds.), *Science, values and objectivity* (pp. 127–142). Pittsburgh: University of Pittsburgh Press.

Losee, J. (2001). *A historical introduction to the philosophy of science.* Oxford: Oxford University Press.

Machamer, P., Pera, M., & Baltas, A. (2000). Scientific controversies: An introduction. In P. Machamer, M. Pera, and A. Baltas (Eds.), *Scientific controversies: Philosophical and historical perspectives* (pp. 3–17). New York: Oxford University Press.

Machamer, P., & Wolters, G. (2004). Introduction. In P. Machamer and G. Wolters (Eds.), *Science, values and objectivity* (pp. 1–13). Pittsburgh: University of Pittsburgh Press.

Margenau, H. (1950). *The nature of physical reality.* New York: McGraw-Hill.

McCann, H. G. (1978). *Chemistry transformed: The paradigmatic shift from phlogiston to oxygen.* Norwood: Ablex.

McComas, W. F. (2008). Seeking historical examples to illustrate key aspects of nature of science. *Science and Education, 17,* 249–263.

McComas, W. F., Almazroa, H., & Clough, M. P. (1998). The role and character of the nature of science in science education. *Science and Education, 7,* 511–532.

McMurry, J., & Fay, R. C. (1998). *Chemistry* (2nd ed.). Upper Saddle River: Prentice Hall.

Millar, R., & Osborne, J. F. (1998). *Beyond 2000: Science education for the future.* London: King's College London.

Mortimer, C. E. (1983). *Chemistry* (5th ed.). Belmont: Wadsworth.

Musgrave, A. (1976). Why did oxygen supplant phlogiston? Research programmes in the chemical revolution. In C. Howson (Ed.), *Method and appraisal in the physical sciences: The critical background to modern science, 1800–1905* (pp. 181–209). Cambridge: Cambridge University Press.

Nagel, E. (1961). *The structure of science.* New York: Harcourt, Brace and World.

National Research Council, NRC. (1996). *National science education standards.* Washington, DC: National Academy Press.

National Society for the Study of Education. (1947). *Science education in American schools: Forty-sixth yearbook of the NSSE.* Chicago: University of Chicago Press.

Niaz, M. (1998). From cathode rays to alpha particles to quantum of action: A rational reconstruction of structure of the atom and its implications for chemistry textbooks. *Science and Education, 82*, 527–552.

Niaz, M. (2000a). The oil drop experiment: A rational reconstruction of the Millikan–Ehrenhaft controversy and its implications for chemistry textbooks. *Journal of Research in Science Teaching, 37*, 480–508.

Niaz, M. (2000b). A rational reconstruction of the kinetic molecular theory of gases based on history and philosophy of science and its implications for chemistry textbooks. *Instructional Science, 28*, 23–50.

Niaz, M. (2001a). How important are the laws of definite and multiple proportions in chemistry and teaching chemistry?—A history and philosophy of science perspective. *Science and Education, 10*, 243–266.

Niaz, M. (2001b). A rational reconstruction of the origin of the covalent bond and its implications for general chemistry textbooks. *International Journal of Science Education, 23*, 623–641.

Niaz, M. (2001c). Understanding nature of science as progressive transitions in heuristic principles. *Science Education, 85*, 684–690.

Niaz, M. (2005a). An appraisal of the controversial nature of the oil drop experiment: Is closure possible? *British Journal for the Philosophy of Science, 56*, 681–702.

Niaz, M. (2005b). The quantitative imperative vs the imperative of presuppositions. *Theory and Psychology, 15*, 247–256.

Niaz, M. (2008a). What 'ideas-about-science' should be taught in school science? A chemistry teachers' perspective. *Instructional Science, 36*, 233–249.

Niaz, M. (2008b). *Teaching general chemistry: A history and philosophy of science approach.* New York: Nova Science Publishers.

Niaz, M. (2009). *Critical appraisal of physical science as a human enterprise: Dynamics of scientific progress.* Dordrecht: Springer.

Niaz, M. (2010). Are we teaching science as practiced by scientists? *American Journal of Physics, 78*(1), 5–6.

Niaz, M. (2011). *Innovating science teacher education: A history and philosophy of science perspective.* New York: Routledge.

Niaz, M., & Cardellini, L. (2011). What can the Bohr–Sommerfeld model show students of chemistry in the 21st century? *Journal of Chemical Education, 88*(2), 240–243.

Niaz, M., & Coştu, B. (2009). Presentation of atomic structure in Turkish general chemistry textbooks. *Chemistry Education Research and Practice, 10*, 233–240.

Niaz, M., & Fernández, R. (2008). Understanding quantum numbers in general chemistry textbooks. *International Journal of Science Education, 30*, 869–901.

Niaz, M., Klassen, S., McMillan, B., & Metz, D. (2010a). Reconstruction of the history of the photoelectric effect and its implications for general physics textbooks. *Science Education, 94*, 903–931.

Niaz, M., Klassen, S., McMillan, B., & Metz, D. (2010b). Leon Cooper's perspective on teaching science: An interview study. *Science and Education, 19*, 39–54.

Niaz, M., & Rodríguez, M. A. (2001). Do we have to introduce history and philosophy of science or is it already 'inside' chemistry? *Chemistry Education: Research and Practice in Europe, 2*, 159–164.

Niaz, M., & Rodríguez, M. A. (2005). The oil drop experiment: Do physical chemistry textbooks refer to its controversial nature? *Science and Education, 14*, 43–57.

Niaz, M., Rodríguez, M. A., & Brito, A. (2004). An appraisal of Mendeleev's contribution to the development of the periodic table. *Studies in History and Philosophy of Science, 35*, 271–282.

Ouellette, R. J. (1975). *Introductory chemistry* (2nd ed.). New York: Harper and Row.

Padilla, K., & Furio-Mas, C. (2008). The importance of history and philosophy of science in correcting distorted views of 'amount of substance' and 'mole' concepts in chemistry teaching. *Science and Education, 17*, 403–424.

Páez, Y., & Niaz, M. (2008). Naturaleza, historia, y filosofía de la ciencia: Un análisis de la imagen reflejada por los textos de química de noveno grado en Venezuela. *Revista de Educación en Ciencias, 9*, 28–31.

Perrin, C. E. (1988). The chemical revolution: Shifts in guiding assumptions. In A. Donovan, L. Laudan, and R. Laudan (Eds.), *Scrutinizing science: Empirical studies of scientific change* (pp. 105–124). Dordrecht: Kluwer.

Peters, E. I. (1990). *Introduction to chemical principles* (5th ed.). Philadelphia: Saunders.

Petrucci, R. H. (1989). *General chemistry: Principles and modern applications* (5th ed.). New York: Macmillan.

Petrucci, R. H., Harwood, W. S., & Herring, F. G. (2003). *General chemistry: Principles and modern applications* (8th ed.). San Francisco: Benjamin Cummings (Pearson Education).

Quagliano, J. V., & Vallarino, L. M. (1969). *Chemistry* (3rd ed.). Englewood Cliffs, NJ: Prentice Hall.

Rodríguez, M. A., & Niaz, M. (2002). How in spite of the rhetoric, history of chemistry has been ignored in presenting atomic structure in textbooks. *Science and Education, 11*, 423–441.

Rodríguez, M., & Niaz, M. (2004a). A reconstruction of structure of the atom and its implications for general physics textbooks. *Journal of Science Education and Technology, 13*, 409–424.

Rodríguez, M. A., & Niaz, M. (2004b). The oil drop experiment: An illustration of scientific research methodology and its implications for physics textbooks. *Instructional Science, 32*, 357–386.

Sadler, T. D., Chambers, F. W., & Zeidler, D. L. (2004). Student conceptualizations of the nature of science in response to a socio-scientific issue. *International Journal of Science Education, 26*, 387–409.

Schwartz, R., & Lederman, N. (2008). What scientists say: Scientists' views of nature of science and relation to science context. *International Journal of Science Education, 30*, 727–771.

Shapin, S. (1996). *The scientific revolution*. Chicago: University of Chicago Press.

Shiland, T. W. (1997). Quantum mechanics and conceptual change in high school chemistry textbooks. *Journal of Research in Science Teaching, 34*, 535–545.

Silberberg, M. S. (2000). *Chemistry: The molecular nature of matter and change* (2nd ed.). New York: McGraw Hill.

Sisler, H. H., Dresdner, R. D., & Mooney, W. T. (1980). *Chemistry: A systematic approach*. New York: Oxford University Press.

Skoog, G. (2005). The coverage of human evolution in high school biology textbooks in the 20th century and in current state science standards. *Science and Education, 14*, 395–422.

Slabaugh, W. H., & Parsons, T. D. (1976). *General chemistry* (3rd ed.). New York: Wiley.

Smith, M. U., & Scharmann, L. C. (1999). Defining versus describing the nature of science: A pragmatic analysis for classroom teachers and science educators. *Science Education, 83*, 493–509.

Smith, M. U., & Scharmann, L. (2008). A multi-year program developing an explicit reflective pedagogy for teaching pre-service teachers the nature of science by ostention. *Science and Education, 17*, 219–248.

Spencer, J. N., Bodner, G. M., & Rickard, L. H. (1999). *Chemistry: Structure and dynamics*. New York: Wiley

Stern, L., & Roseman, J. E. (2004). Can middle-school science textbooks help students learn important ideas? Findings from Project 2061's curriculum evaluation study: Life science. *Journal of Research in Science Teaching, 41*, 538–568.

Stoker, H. S. (1990). *Introduction to chemical principles* (3rd ed.). New York: Macmillan.

Summerlin, L. R. (1981). *Chemistry for the life sciences*. New York: Random House.

Tampakis, C., & Skordoulis, C. (2007). The history of teaching quantum mechanics in Greece. *Science and Education, 16*, 371–391.

Thagard, P. (1990). The conceptual structure of the chemical revolution. *Philosophy of Science, 57*, 183–209.

Toon, E. R., & Ellis, G. L. (1978). *Foundations of chemistry*. New York: Holt, Rinehart Winston.

Tro, N. J. (2008). *Chemistry: A molecular approach.* Upper Saddle River: Prentice Hall (Pearson Education).

Tsai, C.-C. (2007). Teachers' scientific epistemological views: The coherence with instruction and students' views. *Science and Education, 91,* 222–243.

Ucko, D. A. (1982). *Basics for chemistry.* New York: Academic Press.

Umland, J. B., & Bellama, J. M. (1999). *General chemistry* (3rd ed.). Pacific Grove: Brooks/Cole.

Van Berkel, B., DeVos, W., Verdonk, A. H., & Pilot, A. (2000). Normal science education and its dangers: The case of school chemistry. *Science and Education, 9,* 123–159.

Velentzas, A., Halkia, K., & Skordoulis, C. (2007). Thought experiments in the theory of relativity and in quantum mechanics: Their presence in textbooks and in popular science books. *Science and Education, 16,* 353–370.

Wilson, D. (1983). *Rutherford: Simple genius.* Cambridge: MIT Press.

Windschitl, M. (2004). Folk theories of "inquiry:" How preservice teachers reproduce the discourse and practices of an atheoretical scientific method. *Journal of Research in Science Teaching, 41,* 481–512.

Windschitl, M., Thompson, J., & Braaten, M. (2008). Beyond the scientific method: Model-based inquiry as a new paradigm of preference for school science investigations. *Science Education, 92,* 941–967.

Zumdahl, S. S. (1993). *Chemistry* (3rd ed.). Lexington: Heath.

Appendix A
List of General Chemistry Textbooks
Analyzed in this Study (*n* = 75)

Ander, P., & Sonnesa, A. (1965). *Principles of chemistry: An introduction to theoretical concepts.* New York: Macmillan.

Atkins, P., & Jones, L. (1999). *Chemical principles. The quest for insight.* New York: Freeman.

Atkins, P., & Jones, L. (2008). *Chemical principles: The quest for insight* (4th ed.). New York: Freeman.

Bailar, J. C., Jr., Kleinberg, J., Castellion, M., Moeller, T., Guss, C., & Metz, C. (1984). *Chemistry* (2nd ed.). New York: Academic Press.

Bishop, M. (2002). *An introduction to chemistry.* San Francisco: Benjamin Cummings.

Bodner, G. M., & Pardue, H. L. (1989). *Chemistry an experimental science.* New York: Wiley.

Boikess, R., & Edelson, E. (1985). *Chemical principles* (3rd ed.). New York: Harper & Row.

Brady, J. E. (2000). *General chemistry: Principles and structures* (2nd ed., Spanish). New York: Wiley.

Brady, J. E., & Holum, J. R. (1981). *Fundamentals of chemistry.* New York: Wiley.

Brady, J. E., & Holum, J. R. (1996). *Chemistry: The study of matter and its changes* (2nd ed.). New York: Wiley.

Brady, J. E., & Humiston, G. E. (1996). *General chemistry: Principles and structures.* New York: Wiley.

Brady, J. E., Russell, J. L., & Holum, J. R. (2000). *Chemistry: Matter and its changes* (3rd ed.). New York: Wiley.

Brescia, F., Arents, J., Meislich, H., & Turk, A. (1975). *Fundamentals of chemistry.* New York: Academic Press.

Brown, T. L., & Le May, H. E. (1985). *Chemistry: The central science* (3rd ed.). Englewood Cliffs: Prentice Hall.

Brown, T. L., Le May, H. E., Bursten, B. E., & Burdge, J. R. (2003). *Chemistry: The central science* (9th ed.). Englewood Cliffs. Prentice Hall

Burns, R. A. (1995). *Fundamentals of chemsitry* (2nd ed., Spanish). Englewood Cliffs. Prentice Hall.

Chang, R. (1994). *Chemistry* (4th ed., Spanish). New York: McGraw Hill.

Chang, R. (2003). *Chemistry* (7th ed., Spanish). New York: McGraw Hill.

Daub, G. W., & Seese, W. S. (1996). *Basic chemistry* (7th ed., Spanish). New York: Prentice Hall.

Dickerson, R. E., Gray, H. B., Darensbourg, M. Y., & Darensbourg, D. J. (1984). *Chemical principles* (4th ed.). Menlo Park: Benjamin Cummings.

Dickerson, R. E., Gray, H. B., & Haight, G. P. (1974). *Chemical principles* (2nd ed., Spanish). Menlo Park: Benjamins Cummings.

M. Niaz and A. Maza, *Nature of Science in General Chemistry Textbooks,*
SpringerBriefs in Education, DOI: 10.1007/978-94-007-1920-0,
© Mansoor Niaz 2011

Dickson, T. R. (2000). *Introduction to chemistry* (8th ed.). New York: Wiley.

Ebbing, D. D., & Wrighton, M. S. (1993). *General chemistry* (4th ed.). New York: Houghton Mifflin.

Garland, J. K. (1975). *Chemistry of our world*. New York: Macmillan.

Gillespie, R. J., Humphreys, D. A., Baird, N. C., & Robinson, E. A. (1989). *Chemistry* (2nd ed.). Boston: Allyn and Bacon.

Goates, J. R., Ott, J. B., & Butler, E. A. (1981). *General chemistry: Theory and description*. New York: Harcourt Brace Jovanovich.

Goldberg, D. E. (2001). *Fundamentals of chemistry* (3rd ed.). New York: McGraw Hill.

Gray, H. B., & Haight, G. P. (1969). *Basic principles of chemistry* (Spanish ed.). New York: Benjamin.

Hein, M. (1990). *Foundations of college chemistry* (7th ed.). Belmont: Brooks/Cole.

Henold, K. L., & Walmsley, F. (1984). *Chemical principles, properties, and reactions*. Reading: Addison-Wesley.

Hill, J. W. (1975). *Chemistry for changing times* (2nd ed.). Minneapolis: Burgess Publishing Company.

Hill, J. W., & Feigl, D. M. (1987). *Chemistry and life: An introduction to general, organic and biological chemistry* (3rd ed.). New York: Macmillan.

Hill, J. W., & Petrucci, R. H. (1999). *General chemistry: An integrated approach* (2nd ed.). Upper Sadle River: Prentice Hall.

Holum, J. R. (2000). *Fundamentals of general, organic and biochemistry for the health sciences*. New York: Wiley.

Joesten, M. D., Johnston, D. O., Netterville, J. T., & Wood, J. L. (1991). *World of chemistry*. Philadelphia: Saunders.

Jones, L., & Atkins, P. (2000). *Chemistry: Molecules, matter and changes* (4th ed.). New York: Freeman.

Keenan, C. W., Kleinfelter, D. C., & Wood, J. H. (1985). *General college chemistry* (6th ed.). New York: Harper & Row.

Kotz, J. C., & Treichel, P. (1999). *Chemistry and chemical reactivity* (4th ed.). Philadelphia: Saunders.

Lippincott, W. T., Garrett, A. B., & Verhoek, F. H. (1977). *Chemistry: A study of matter* (3rd ed.). New York: Wiley.

Mahan, B., & Meyers, R. J. (1990). *University chemistry* (4th ed.). Wilmington: Addison-Wesley.

Malone, L. J. (2001). *Basic concepts of chemistry* (6th ed.). New York: Wiley.

Masterton, W. L., & Slowinski, E. J. (1980). *Chemical principles* (3rd ed.). New York: McGraw Hill.

Masterton, W. L., Slowinski, E. J., & Stanitski, C. C. (1985). *Chemical principles* (6th ed.). New York: McGraw Hill.

Matta, M. S., & Wilbraham, A. C. (1981). *Atoms molecules and life. An introduction to general organic and biological chemistry*. Menlo Park: Benjamin Cummings.

McMurry, J., & Fay, R. C. (1998). *Chemistry* (2nd ed.). Upper Saddle River: Prentice Hall.

Miller, F. M. (1984). *Chemistry: Structure and dynamics*. New York: McGraw Hill.

Moore, J. W., Davies, W. G., & Collins, R. W. (1978). *Chemistry*. New York: McGraw Hill.

Moore, J. W., Stanitski, C. L., & Jurs, P. C. (2002). *Chemistry: The molecular science*. Orlando: Harcourt College.

Mortimer, C. E. (1983). *Chemistry* (5th ed.). Belmont: Wadsworth.

Murphy, D. B., & Rousseau, V. (1980). *Foundations of college chemistry* (3rd ed.). New York: Wiley.

Ouellette, R. J. (1975). *Introductory chemistry* (2nd ed.). New York: Harper & Row.

Oxtoby, D. W., Gillis, H. P., & Nachtrieb, N. H. (1999). *Principles of modern chemistry* (4th ed.). Philadelphia: Saunders.

Oxtoby, D. W., Nachtrieb, N. H., & Freeman, W. A. (1990). *Chemistry: Science of change* (2nd ed.). Philadelphia: Saunders.

Peters, E. I. (1990). *Introduction to chemical principles* (5th ed.). Philadelphia: Saunders.

Petrucci, R. H. (1989). *General chemistry: Principles and modern applications* (5th ed.). New York: Macmillan.

Petrucci, R. H., Harwood, W. S., & Herring, F. G. (2003). *General chemistry: Principles and modern applications* (8th ed.). San Francisco: Benjamin Cummings (Pearson Education).

Phillips, J. S., Strozak, V. S., & Wistrom, C. (2000). *Chemistry: Concepts and applications.* New York: McGraw Hill.

Quagliano, J. V., & Vallarino, L. M. (1969). *Chemistry* (3rd ed.). Englewood Cliffs: Prentice Hall.

Russo, S., & Silver, M. (2002). *Introductory chemistry* (2nd ed.). San Francisco: Benjamin Cummings.

Segal, B. G. (1989). *Chemistry: Experiment and theory* (2nd ed.). New York: Wiley.

Sherman, A., Sherman, S. J., & Russikoff, L. (1992). *Basic concepts of chemistry* (5th ed.). Boston: Houghton Mifflin.

Silberberg, M. S. (2000). *Chemistry: The molecular nature of matter and change* (2nd ed.). New York: McGraw Hill.

Sisler, H. H., Dresdner, R. D., & Mooncy, W. T. (1980). *Chemistry: A systematic approach.* New York: Oxford University Press.

Slabaugh, W. H., & Parsons, T. D. (1976). *General chemistry* (3rd ed.). New York: Wiley.

Spencer, J. N., Bodner, G. M., & Rickard, L. H. (1999). *Chemistry: Structure and dynamics.* New York: Wiley.

Stoker, H. S. (1990). *Introduction to chemical principles* (3rd ed.). New York: Macmillan.

Summerlin, L. R. (1981). *Chemistry for the life sciences.* New York: Random House.

Toon, E. R., & Ellis, G. L. (1978). *Foundations of chemistry.* New York: Holt, Rinehart & Winston.

Tro, N. J. (2008). *Chemistry: A molecular approach.* Upper Saddle River: Prentice Hall (Pearson education).

Ucko, D. A. (1982). *Basics for chemistry.* New York: Academic Press.

Umland, J. B., & Bellama, J. M. (1999). *General chemistry* (3rd ed.). Pacific Grove: Brooks/Cole.

Whitten, K. W., Davis, R. E., & Peck, M. L. (1996). *General chemistry* (5th ed., Spanish). New York: McGraw Hill.

Williams, A. L., Embree, H. D., & De Bey, H. J. (1981). *Introduction to chemistry* (3rd ed.). Reading: Addison-Wesley.

Wolfe, D. H. (1988). *Introduction to college chemistry* (2nd ed.). New York: McGraw Hill.

Zumdahl, S. S. (1993). *Chemistry* (3rd ed.). Lexington: Heath.

Appendix B
Reliability of Evaluation of Textbooks Based on Inter-Rater Agreement

Criteria	Agreement (n = 75 texts)	Percentage (%)
1	65	86.7
2	71	94.7
3	67	89.3
4	72	96.0
5	68	90.7
6	73	97.3
7	70	93.3
8	69	92.0
9	72	96.0
Average		92.9

M. Niaz and A. Maza, *Nature of Science in General Chemistry Textbooks*,
SpringerBriefs in Education, DOI: 10.1007/978-94-007-1920-0,
© Mansoor Niaz 2011